ALUMINIUM MATERIALS TECHNOLOGY FOR AUTOMOBILE CONSTRUCTION

LIST OF AUTHORS

Dr-Ing. F. Ostermann
Vorstand
Aluminium-Zentrale e.V.
Düsseldorf

Dr rer. nat. A. Blecher
VAW
Vereinigte Aluminium-Werke AG
Bonn

Dipl.-Ing. D. Brandt
Otto Fuchs Metallwerke
Meinerzhagen

Dr-Ing. B. Leuschen
Mercedes-Benz AG
Verfahrensentwicklung
Sindelfingen

Dr-Ing. H. Lowak
Oyyo Fuchs Metallwerke
Meinerzhagen

Dr rer. nat. Jorg Maier
ALUSINGEN GmbH
Abt. ZLW
Singen/Hohentwiel

Dr rer. nat. Frank Wehner
ALUSINGEN GmbH
Singen/Hohentwiel

Dipl.-Ing. Willi Wurl
Dr Ing. h. c. F. Porsche AG
Abt EKV 1
Weissach

Dipl.-Ing. K.-H. v. Zengen
Vereinigte Aluminium-Werke AG
Bonn

Aluminium Materials Technology for Automobile Construction

Dr-Ing. Friedrich Ostermann

Dr rer. nat. A. Blecher
Dipl.-Ing. W. Wurl
Dr-Ing. B. Leuschen
Dr rer. nat. J. Maier
Dr rer. nat. F. Wehner
Dipl.-Ing. K.-H. v. Zengen
Dr-Ing. H. Lowak
Dipl. Ing. D. Brandt

English translation edited by Roy Woodward

Translated from the German by Pam Chatterley, BA, MITI

Mechanical Engineering Publications Limited
LONDON

First published (in German) 1992
© Expert Verlag, 7044 Ehningen bei Böblingen

English language Edition 1993
© Mechanical Engineering Publication Limited

ISBN 0 85298 880 X

A CIP catalogue record for this book is available from the
British Library.

Typeset by Santype International Limited,
Netherhampton Road, Salisbury, Wiltshire
Printed by Lavenham Press, Lavenham, Suffolk

CONTENTS

TRANSLATION EDITOR'S FOREWORD

The automobile and aluminium became commercially viable at about the same time in the late years of the 19th Century; there are references to the use of the latter in the former from their very beginnings!

In the past twenty years the use of large amounts of aluminium in mass-produced cars – as distinct from expensive low-volume models – has been frequently predicted but, as yet, has not come about. Simply stated, it is proven that aluminium can be used: to replace steel for wheels, body panels, space frames, and bumpers; as a substitute for cast iron in engine blocks, sump covers, and the like; to replace copper in radiators and heat exchangers. In all cases weight savings result with no reduction in performance, but in most instances there is a significant increase in cost. That increase can be countered – on grounds of reduced fuel consumption, increased ability to carry the equipment needed to reduce pollution and improve passenger safety, and increased life of the car – if the user, the manufacturer, and, perhaps most importantly, the legislators, deem them of sufficient merit.

While these debates continue the aluminium producers and the car manufacturers have worked together to narrow the cost gap by: (1) seeking to improve alloy performance to keep the quantity of what is a relatively expensive material used to a minimum, (2) improving designs to make full use of the wide variety of forms in which aluminium is readily available, (3) improving the already efficient techniques of casting, joining, and surface treatment to further reduce fabrication costs, and (4) improving the fundamental understanding of the factors influencing such issues as formability, adhesion, fatigue, corrosion, and so on.

The nine chapters of this book describe, in considerable detail, the progress made in recent years by some German workers, with ample acknowledgement, in 135 references, to the work of others in the UK, the USA, and Japan. It will serve as a valuable input to the deliberations of those concerned with, or contemplating, the use of aluminium for cars and car components. It should also be valuable to Universities and other teaching establishments involved in automobile engineering, since it provides, in one cover, a great deal of the detail about aluminium which is often lacking in the libraries of such institutions.

While it can be argued that additional difficulties will be encountered when aluminium is widely used in greater quantities than is the case today, the importance of the market to the aluminium industry, and of the material to the car industry is such that any problems will certainly receive the attention necessary for their solution; to have come so far and not to succeed would be unthinkable!

Roy Woodward
August 1993

AUTHOR'S FOREWORD

Since 1886 – the start of both the automobile age and of the industrial production of aluminium – the vehicle designer has had a keen interest in the application of the lightweight metal. Today, more than 100 years later, almost a quarter of the aluminium produced worldwide is used in vehicle construction, making this its most significant application. Nevertheless, the amount of aluminium in a vehicle has remained limited to about 3–5 percent (by weight), concentrated mainly in the area of the power units and assemblies.

In spite of the intensive attempts of vehicle builders to achieve lightweight designs with lightweight materials – especially since the energy crises of the 1970s – vehicles have continued to become heavier. Improved occupant protection, reduced noise and vibration, exhaust purification equipment, and increasing vehicle size have negated the effects of weight reduction.

Further noticeable weight reductions will only be achieved with the rigorous application of light construction materials in the body and chassis. Specialists in the automotive industry consider that the aluminium body has the most promising prospects. Not only has the automotive suitability and safety of aluminium bodies and components been demonstrated over many years of development, but also the value retained by the alloys used following a complete and economic recycling process make them extremly attractive for the future.

Since 1989 the 'Aluminium materials technology for vehicle construction' Seminars at the Esslingen Technical Academy have been devoted specifically to disseminating information on developments in aluminium for this application. This volume is based on the experiences of these seminars, and provides a concentrated overview of current knowledge in contributions from materials specialists, designers, and production engineers.

I would like to thank my co-authors for providing their specialised knowledge in papers, which – as so often – have been produced in their own time, as well as all others who have supported the production of this book.

Friedrich Ostermann
Mechenheim, October 1991

GLOSSARY OF TERMS AND DESIGNATIONS FOR CASTING ALLOYS

DIN	EN*	Description
G-	-s	Sand casting
GK-	-k	Chill or permanent mould casting
GD-	-D	Pressure diecasting
GF-	-L	Investment casting
-g	-O	Annealed
-ka	-T4	Solution heat treated and naturally aged
-wa	-T6	Solution heat treated and artificially aged
-ta	-T64	Solution heat treated and artificially under-aged
-dv	none	Grain refinement by strontium addition

* Proposed designations in new European Standards, according to document PREN 132/100 1992, 5th Draft.

Aluminium materials for vehicle construction

F. Ostermann

1.1 LIGHTWEIGHT CONSTRUCTION USING ALUMINIUM

The chassis, power unit, and body are the main component areas of a vehicle. In addition there is the equipment area with an increasing proportion of heavy comfort equipment (see Fig. 1.1) (1). Vehicle manufacturers have a long tradition of attempting to achieve lightweight designs in all the assembly areas mentioned, and – due especially to the energy crises of the 1970s – this has reached a high level of development. Weight reduction by using specifically lighter materials has also been pursued intensively, taking into account what is economically worthwhile. Plastic has profited not least from this. In the past decade and a half the proportion of plastic in the modern vehicle has reached over eight per cent due to intensive efforts on the part of the automotive and plastics industries.

The use of aluminium as a material for vehicles also has a long tradition. In the early decades of this century all-aluminium vehicles were built in small production runs, in order to achieve the best possible performance with the low outputs of the engines of the time. In this connection we should also remember the five-seat Dyna Panhard 1954, which achieved a maximum speed of 125 km/h and a fuel consumption of less than 6 l/100 km at 90 km/h in all-aluminium construction with an 850 cc engine. The basic body, consisting of a tubular chassis frame and a spot-welded body made from $AlMg_3$ sheet, weighed only 98 kg in all, including doors and bonnet (2).

Nevertheless, a vehicle built in Germany today only has an average 50 to 60 kg aluminium content, or about 5 per cent of the vehicle weight. Table 1.1 shows the average distribution of aluminium parts among the various component areas of the vehicles. Most is used in the transmission and engine, followed by the chassis. Tables 1.2(a) to 1.2(e) contain numerous examples of applications for individual components from production vehicles and a distinction is made between finished castings and

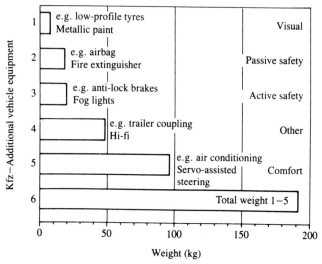

Fig. 1.1   Weight of vehicle add-on equipment, based on an upper mid-range vehicle (1)

semi-finished products. The impression is given that the suitability of the material for vehicle uses and its versatility are undisputed.

With regard to the chassis, aluminium wheels, which have now achieved a market share of about 25 per cent, account for the largest proportion of the weight. It is true that light-alloy wheels have con-

Table 1.1   The use of aluminium in vehicles, 1989

Average aluminium content of car by weight: ca. 5 per cent
Average content of cast and wrought alloys

Component groups	Distribution of Al amongst the groups (%)	(kg)	Cast Al alloys (%)	Wrought Al alloys (%)
Chassis	30	16	90	10
Engine/ transmission	50	28	100	0
Body	15	8	20	80
Equipment	5	3	40	60
Total	100	55	85	15

Table 1.2(a)   **Finished aluminium castings in production vehicles**

Examples	Die cast	Chill cast	Sand cast	Typical weight vehicle in kg
Engine/drivetrain				
piston		X		1–3.5
cylinder head		X		10–20
rocker cover	X	X		1–2.5
engine block		X		14–45
engine cover	X			1–3
oil sump	X	X	X	3–5
induction system		X	X	2–7
gearbox casing	X			8–15
injection system/				
carburettor	X			1–3
water pump casing	X			0.5
petrol pump casing	X			0.5
oil pump casing	X			0.5
engine block mount	X	X		1.5–2.5
engine bearing	X			0.5
differential casing	X	X	X	5–8
Steering/chassis				
steering box	X	X		1.5–2.5
steering wheel	X			0.5
pedal bracket etc.	X			1–3
wheels		X		27–35
brake master cylinder		X		0.5
ABS		X		1
Body				
window lifter	X			1–3
mirror housing	X			1–1.5
roof rail supports	X	X		1.5–2.5

quered the market not because of weight, but for styling reasons. Functional, really light light-alloy wheels were successfully developed and partially implemented in the 1970s, but they did not achieve any success.

The reasons for lack of success in the application of aluminium in automotive construction are obvious:

– higher material costs;
– higher processing costs;
– inadequate development of materials technology, particularly in joining technology;

Table 1.2(b)   Sample applications of semi-finished aluminium products in production vehicles. Component group: engine, drivetrain

Component	Type of semi-finished product	Units/ vehicle	Weight/ vehicle (kg)
Engine			
engine mount	Impact extrusion	2	0.3
oil cooler	Sheet	1	0.5
oil filter housing	Impact extrusion	1	0.2
filler neck	Tube	1	0.4
fuel filter housing	Impact extrusion	1	0.1
petrol pump housing	Impact extrusion	1	0.1
radiator	Sheet, tube	1	3.5
air filter housing	Sheet	1	0.3
cover	Sheet	1	0.2
Exhaust system			
heat shield	Sheet	1	0.6
Transmission			
gear selector fork	Forging	1	0.2
gear lever	Forging	1	0.2
ring bearing	Forging	1	0.2
flange, bearing outlet	Forging	1	0.2
Injection			
fuel distributor	Forging	1	0.5
plunger	Impact extrusion	1	0.1

– inadequate knowledge and experience among designers and production engineers with regard to using aluminium in large-scale production.

While aluminium has been able largely to conquer the drivetrain and heat exchanger areas, the chassis, body and equipment must be regarded as 'development areas' for lightweight construction using aluminium. The development aims are worthwhile: by using aluminium to build the body it is possible to save about 150 kg of weight directly, which does not include the secondary weight savings. Such a step leads to noticeable reductions in fuel consumption, as can be seen in Fig. 1.2 (3). Extended to the vehicle population of the former Federal Republic of Germany

Table 1.2(c) Sample applications of semi-finished aluminium products in production vehicles. Component group: body

Component	Type of semi-finished product	Units/ vehicle	Weight/ vehicle (kg)
Bonnet	Sheet	1	4.0
reinforcement	Sheet	1	2.8
Boot lid	Sheet	1	3.2
reinforcement	Sheet	1	2.5
Wings	Sheet	2	7.6
Doors			
facing panel	Sheet	4	9.2
stiffening	Sheet	4	10.0
equipment bracket	Sheet	4	7.8
impact member	Extrusion	4	4.0
door hinge	Forging	8	0.5
Window			
window frame	Extrusion	4	3.5
trim strip	Extrusion, strip	6	1.5
guide rail	Extrusion	4	0.2
Sliding/hinged sun-roof			
frame	Extrusion	1	1.4
Roof rail	Extrusion	1	1.5
Roof drip moulding	Extrusion	2	0.3
Hardtop (for roadster)	Sheet	1	–

(over 30 million vehicles), fuel consumption would fall by a significant 3 billion litres per year, and CO_2 emissions would fall accordingly by 7.5 million tonnes/year (see Fig. 1.3) (**4**).

Another area is becoming increasingly important, namely the economic disposal of used vehicles using materials recycling. Aluminium offers by far the best prospects here. About 95 per cent of the aluminium in old vehicles is already regained. New technologies will increase this percentage even further and will facilitate the classification of alloys by grade. The low energy requirement for processing the scrap into new semi-finished products is an important contribution to solving ecological problems.

Table 1.2(d)   Sample applications for semi-finished aluminium products in production vehicles. Component group: chassis, wheels

Component	Type of semi-finished product	Units/ vehicle	Weight/ vehicle (kg)
Wheels	Forging	4	25.2
Suspension			
swivel bearing	Forging	2	3.2
lateral control arm	Forging	2	3.6
radius arms (multi-link)	Extrusion	–	–
Brakes			
brake piston	Impact extrusion	4	0.1
brake servo	Sheet	1	0.5
brake proportioning unit	Extrusion	1	0.4
ABS housing	Forging	1	1.5
ABS bolted connection	Impact extrusion	2	0.1
Steering			
articulated fork	Extrusion	2	0.6
connector piece	Forging	1	0.6
steering wheel	Tube, rod	1	–
Bumpers (where fitted)	Extrusion	2	9.8
brackets	Extrusion	4	0.2

1.2 DEVELOPMENT VEHICLES AND NEW CONCEPTS

During the past two decades various vehicle manufacturers have repeatedly attempted to assess the status of aluminium materials technology by building development vehicles. New types of alloys and advanced production techniques have been tested. Interest has been focussed mainly on testing suitable joining methods, however, as the list in Table 1.3 shows.

The Honda NS-X is the only all-aluminium vehicle which is made in a limited production run. To what extent it can be regarded as a showpiece for aluminium materials technology in vehicle construction is not yet known.

A high level of activity in aluminium can currently be detected at a

Table 1.2(e)   Sample applications for semi-finished aluminium products in production vehicles. Component: equipment, trim

Component	Type of semi-finished product	Units/ vehicle	Weight/ vehicle (kg)
Instrument panel	Sheet	1	1.1
Seats, front			
frame and rail	Sheet, extrusion	2	3.2
backrest	Sheet, tube	2	1.6
head rest	Sheet	4	0.8
seat shell	Sheet	2	3.2
Seats, rear			
frame	Sheet	1	1.5
backrest	Sheet	1	1.4
armrest	Sheet	1	0.4
Air bag gas generator	Sheet	1	0.2
generator mounting	Sheet	1	0.2
Belt tensioner	Sheet, impact extrusion	4	0.1
Air-conditioning system			
condenser	Sheet, tube	1	1.2
vaporiser	Sheet, tube	1	0.8
pipes	Tube	1	0.1
Heater core	Sheet, tube	1	1.1
Jack	Sheet, extrusion	1	1.1
Number plates	Sheet	2	0.3
Identification plates	Sheet	2	0.1

number of vehicle manufacturers, the aim being to find satisfactory solutions to the body-building problems mentioned. The development paths being followed can be approximately divided into three different approaches.

(1)   The development of forming and joining methods and of developing aluminium materials with the aim of reducing the technology and cost gap between aluminium and steel.
(2)   The development of a materials concept based on specially pre-treated grades of sheet and an appropriate joining method, suitable

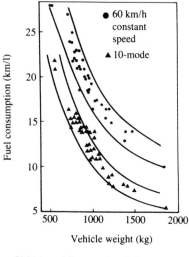

Vehicle weight = curb weight +110 kg
*1985 data from JAMA

Fig. 1.2   Relationship between fuel consumption and vehicle weight (3)

for large-scale production, while maintaining the proven mono-coque body design made of sheet components, and thus the continued use of advanced sheet body production plant.

(3)   Comprehensive development of independent concepts for designs and production methods which make optimum use of the particular advantages of aluminium in forming by extrusion and casting and adapting them to the requirements of vehicle construction.

All three approaches are being used intensively at present. It is to be expected that successful results in all three areas of development will be reflected in any significant breakthrough.

From the point of view of economic efficiency, in particular, three manufacturing techniques, which offer particular advantages for aluminium products, stand out:

– strip coating processes;
– extrusion processes;
– mould casting processes.

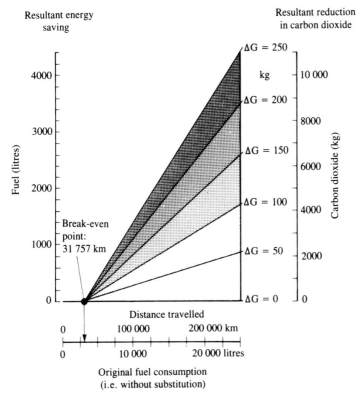

Resultant energy saving

Resultant reduction in carbon dioxide

Fuel (litres)

Carbon dioxide (kg)

Break-even point: 31 757 km

Distance travelled

Original fuel consumption (i.e. without substitution)

Fig. 1.3   The effect of the proportion of aluminium in vehicle weight in reducing fuel consumption and CO_2 emissions (4)

Strip coating is suitable for the construction of sheet monocoque bodies insofar as specifically pre-treated grades of sheet can be produced which can provide:

– favourable tribological characteristics for the production of sheet body parts;
– a reliable adhesive bonding technique for the bodyshell;
– better handling during use;
– savings in pre-treatment.

The deliberate exploitation of these advantages is the subject of the ASVT concept (Alcan's Aluminium Structured Vehicle Technology) (**10,**

Table 1.3   Aluminium development vehicles since 1973

Project/Year Company	Weight saving (%)	Material/Type of semi-finished product	Fastening method used
Corvette/1975 **(5)** GM–Reynolds	39	AA2036, AA5182 Sheet	Spot weld adhesive bonding
Audi 100/1981 Audi, ALUSINGEN, Alcoa	45	AA6016 Sheet	Spot weld adhesive bonding Spot welding Adhesive bonding
Porsche 928 **(6)(7)** Porsche, ALUSINGEN	47	AA6016, AA5182 Sheet	Spot welding Fusion welding
LCP 2000/ca. 1983 Volvo	45	Type AlMgSi Sheet	
Audi 100/1985 **(8)** Audi, Alcoa	47	AA6009 Sheet	Electromagnetic rivetting Adhesive bonding
ECV 3/1983 **(9)(10)** British Leyland, Alcan	45	AA5252, AA5454 Sheet	Pretreated strip Adhesive bonding Spot weld adhesive bonding
Metro/ca.1986 **(10)** Austin Rover, Alcan	46		
X1/9 Cabrio/ca.1986 **(10)** Bertone, Alcan	33		
Fiero/ca.1986 **(10)** GM Pontiac, Alcan	45		
Urban car study/1987 **(11)** Daimler-Benz AG		G-AlSiMg7 wa and wrought Al alloy	Bolting of large surface-area castings
Limited series/1987 **(12)** Treser, Hydro Aluminium		Extrusions Castings	Rivetting Adhesive bonding
Probe V/1968 Ford USA, Reynolds		Extrusions Type AlMgSi	Fusion welding of space frame
'AZ550 Sports'/1988 **(13)** Mazda		Extrusions with honeycomb panels	Welded and brazed
NS–X Sportscar/1990 Honda	45	Sheet, extrusions	Spot welded, MIG welded

14–16). Some European vehicle builders have built development vehicles using this materials technique and they appear to have been successfully tested (**17, 18**). The bodyshell of an Austin Rover Metro (Fig. 1.4) made in this way weighed only 74 kg compared with 137 kg in the corresponding steel design **10**).

Making bodies with extruded and mould-cast products, on the other hand, requires new design and production techniques. Aluminium is clearly superior to all other materials in its suitability for the manufacture of thin-walled extruded sections and castings with a high level of styling freedom and high ductility, together with comparatively low tooling costs.

To utilize these advantages, the so-called space-frame design was developed for aluminium bodies (**12, 14, 19, 20**). For reasons of rigidity, hollow sections are primarily used for the frame structure, and joined using adhesive bonding, riveting, and welding processes. Thin-walled mould castings are used for particularly complex frame nodes and frame parts.

Figure 1.5 (**14**) shows an example of such a space-frame structure. A comparison of the estimated tooling costs for conventional sheet bodies and space-frame bodies indicates considerable savings for the latter in small- and medium-sized production runs (see Fig. 1.6) (**14**). Additional advantages include flexibility of design variations, 'modular design' which has cost benefits, and shorter model development times.

Fig. 1.4   Bodyshell of the Austin Rover *Metro* in aluminium (10)

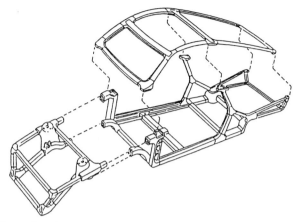

Fig. 1.5   Example of a space-frame design consisting of extruded sections and cast nodes (14)

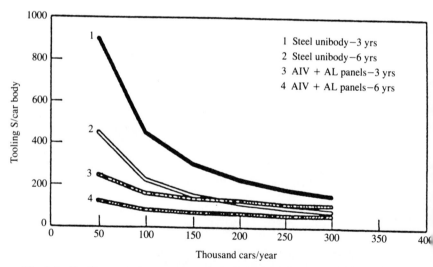

1   Steel unibody−3 yrs
2   Steel unibody−6 yrs
3   AIV + AL panels−3 yrs
4   AIV + AL panels−6 yrs

Fig. 1.6   Tooling costs per vehicle for conventional steel design and for the AIV design (space frame) over a 3-year and a 6-year production cycle (AIV = *A*luminium *I*ntensive *V*ehicle study by Alcoa)

All construction concepts mentioned, including those production concepts based on existing technology, require development directed at the specific needs of vehicle construction, and mastery of the whole field of aluminium materials technology which may be divided into the classical five disciplines:

- materials and their characteristics;
- concept, design and analysis;
- production technology, forming processes;
- joining methods;
- surface engineering.

Below is a survey of aluminium materials which have been specially designed in the rolling sector for vehicle construction or which are suitable for it. The design, manufacturing and production aspects are dealt with in more detail in later chapters.

1.3 SURVEY OF MATERIALS

The development of new alloys and the optimization of standard alloys with improved service and processing characteristics for automotive applications have produced, nationally and internationally, a wide range of aluminium materials, which are not contained, or are contained only in part, in existing materials standards. For this reason it is not possible to give reliable minimum values for composition and characteristics for all alloys. Reference will therefore be restricted to indicating nominal compositions and typical values for service properties, as extracted from relevant company information or technical papers. In view of the energetic development of sheet body materials in Japan, this information will also be taken into account for the sake of interest, so far as it is available.

Table 1.4 surveys the composition of alloys which, with the exception of Al99.5, were developed specifically for bodysheet applications. Table 1.5 contains the corresponding mechanical properties, and Table 1.6 gives an indication of applications of the various alloys.

Group 2 alloys

Magnesium (Mg) is the characteristic alloying component in Group 2 alloys. These alloys are characterized by a very good combination of formability, strength, and corrosion resistance, and are therefore preferred for difficult deep-drawn pieces.

Table 1.4 Nominal composition of Al body sheet alloys

Designation		Nominal composition (wt%)					
DIN 1725	*ISO*	*Si*	*Cu*	*Mn*	*Mg*	*Cr*	*Zn*
Group 1: unalloyed aluminium							
Al99.5	1050	–	Nominal aluminium content 99.5%				–
Group 2: AlMg(Mn,Cr), not precipitation hardenable							
AlMg2.5	5052	–	–	–	2.5	0.2	–
AlMg3	5742	–	–	0.2	3.0	0.1†	–
AlMg5Mn	5182	–	–	0.4	4.5	–	–
Group 3: AlMgSi(Cu,Mn), precipitation hardenable							
–	6009	0.8	0.4	0.5	0.6	–	–
–	6010	1.0	0.4	0.5	0.8	–	–
(AlMg0.4Si1.2)*	6016	1.2	–	–	0.4	–	–
–	6111	0.9	0.7	0.2	0.7	–	–
Group 4: AlCuMg(Si), precipitation hardenable							
–	2002	0.7	2.0	–	0.8	–	–
–	2008	0.7	0.9	–	0.4	–	–
–	2036	–	2.6	0.3	0.5	–	–
–	2038	0.8	1.3	0.3	0.7	–	–
Group 5: AlMgCu(Zn), precipitation hardenable							
–	GZ45/30–30	–	0.4	–	4.5	–	1.5
–	KS5030	–	0.50	–	4.5	–	–

* designation in ( ) not standard in DIN 1725
† at least one of the elements Mn and Cr must be present.

One disadvantage is the occurrence of flow lines on the sheet surface when producing deep-drawn parts. Two types of flow-line markings can be distinguished. Type A flow lines occur as a result of stretching at the start of flow which increases with magnesium content. Type B flow lines are due to strain ageing which increases with plastic stretching (Portevin–Le Chatelier Effect) at certain elongation speeds carried out at room temperature (see Fig. 1.7). Type A stretcher strains can be avoided by means of suitable manufacturing conditions (a combination of appropriately selected grain size, heat treatment and levelling process). These alloys are then said to have low stretcher strain or stretcher-strain-free (ssf) qualities.

The static strength of these alloys is determined by the so-called solid–

Table 1.5 Typical strength properties of Al body sheet alloys

Designation		State	R_m $-(N/mm^2)$	$R_{p0.2}$ $-$	A_5 $-(\%)$	A_{gl} $-$	n	r	I_e^* (mm)	$B_{0\,max}$
DIN	ISO									
Group 1: unalloyed aluminium										
Al99.5 W7	1050–0	Soft	80	40	40	28	0.25	0.85	10.5	2.1
Group 2: AlMg(Mn,Cr), not precipitation hardenable										
AlMg2.5 W18	5052–0	soft	190	90	28	24	0.3	0.68	–	2.1
AlMg3 W19	5754–0	soft	210	100	28	19	0.3	0.75	9.4	2.1
AlMg5Mn W27	5182–0	soft	280	140	30	23	0.31	0.75	10.0	2.1
AlMg5Mn	5182–ssf	soft	270	125	24	–	0.31	0.67	–	–
Group 3: AlMgSi(Cu,Mn), precipitation hardenable										
–	6009–T4	cph†	230	125	27	–	0.23	0.70	–	–
–	6010–T4	cph	290	170	24	–	0.22	0.70	–	–
AlSi1.2Mg0.4	6016–T4	cph	240	120	28	–	0.27	0.65	10.2	2.1
–	6111–T4	cph	275	160	28	–	0.26	0.56	–	–
Group 4: AlCuMg(Si), precipitation hardenable										
–	2002–T4	cph	330	180	26	–	0.25	0.63	9.6	–
–	2008–T4	cph	250	140	28	–	0.28	0.58	–	–
–	2036–T4	cph	340	190	24	–	0.23	0.70	–	–
–	2038–T4	cph	320	170	25	–	0.26	0.70	–	–
Group 5: AlMgCu(Zn), precipitation hardenable										
–	GZ45/30–30	cph	300	155	30	–	0.29	0.68	9.8	–
–	KS5030–T4	cph	275	135	30	28	0.3	0.65	9.8	2.08

* I_E = capping index at 1 mm thickness
† cph = cold precipitation hardened

Table 1.6 Applications of Al body sheet alloys

Designation		Propensity to stretcher strains	Hem type	Application examples
DIN	ISO			
Group 1: Unalloyed aluminium				
Al99.5 W7	1050–0	None	Normal	Heat screening Number plates
Group 2: AlMg(Mn,Cr), not precipitation hardenable				
AlMg2.5 W18	5052–0	{A}*,B	Normal	Interior body parts
AlMg3 W19	5754–0	A,B	Normal	Interior body parts
AlMg5Mn W27	5182–0	A,B	Normal	Interior body parts
AlMg5Mn	5182–ssf	B	Normal	Interior/exterior body parts
Group 3: AlMgSi(Cu,Mn), precipitation hardenable				
–	6009–T4	None	Normal	Interior/exterior body parts
–	6010–T4	None	Rope	Exterior body parts
AlMg0.4Si1.2	6016–T4	None	(Normal)	Exterior body parts
–	6111–T4	None	Normal	Exterior body parts
Group 4: AlCuMg(Si), precipitation hardenable				
–	2002–T4	None	Rope	Exterior body parts
–	2008–T4	None	Rope	Exterior body parts
–	2036–T4	None	Rope	Exterior body parts
–	2038–T4	None	–	Exterior body parts
Group 5: AlMgCu(Zn), precipitation hardenable				
–	GZ45/30–30	{B}	Normal	Interior/exterior body parts
–	KS5030	{B}	Normal	Interior/exterior body parts

* based on assumptions
A,B = type of stretcher strain–see text
ssf = stretcher strain free (Type A)

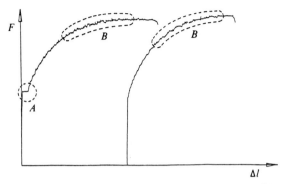

Fig. 1.7   **Force/elongation diagram for AlMg5Mn.** *Left*: 'soft' state; *right*: 'stretcher-strain-free' (ssf) state

solution hardening of the aluminium by magnesium. The high strengthening capacity of this group of alloys during plastic deformation favours the stretch-formability characteristics. It is true that the high strain hardening is partially reversed even at temperatures of 180° to 200°C, so a reduction in strength must be expected with conventional paint firing (see Table 1.7).

Table 1.7   **The effect of cold working and paint curing on the strength characteristics of some body sheet alloys (22)**

Designation	State	R_m (N/mm^2)	$R_{p0.2}$ (N/mm^2)	A_5 (%)
AlMg3 (5754)	Soft	210	90	28
	10% cold worked	235	220	17
	10% cold worked + 190°C/0.5h	235	180	22
AlMg5Mn (5182)	Soft	275	135	30
	10% cold worked	285	260	16
	10% cold worked + 190°C/0.5h	275	185	20
AlSi1.2Mg0.4 (6016)	Cold precipitation hardened	240	120	34
	10% cold worked	270	210	22
	10% cold worked + 190°C/0.5h	290	260	15

For processing it is also important that the oxide layer thickness can increase to about 0.1–0.2 μm with increasing Mg content as the strip is soft annealed. On the one hand, this has a favourable effect on tribological characteristics in forming, in conjunction with appropriate lubrication, but, on the other hand, it is unfavourable for reproducible results in spot welding and adhesive bonding. It is therefore necessary to include a separate pickling degreasing process between forming and joining to achieve optimal manufacturing parameters, unless special strip pre-treatments, such as strip chromating, are to be used.

Finally, it must be pointed out that the AlMg5Mn alloy exhibits a low electrical conductivity (15–19 m/Ωmm^2) compared with all other alloys and can thus be spot welded with lower current strengths.

Group 3 alloys

The characteristic alloying elements of this group are silicon (Si) and magnesium (Mg), with which aluminium forms a precipitation-hardening alloying system. The low Cu content noticeable in American alloy developments also serves to increase strength. A characteristic of all alloys in this group is a certain excess of Si, which ensures fast response of the precipitation-hardening mechanism. The alloys in this group are therefore precipitation-hardenable and must be supplied and processed in the cold precipitation-hardened state (cph or T4).

In contrast to the AlMg alloys in Group 2, the AlMgSi alloys are characterized by freedom from stretcher strain and high strength, which is improved further by the paint firing (see Table 1.7). The preferred application of these alloys is therefore the outer skin area of the vehicle, which must meet high standards of appearance and buckling resistance.

In the case of attached parts, in particular, such as bonnet, boot lid and doors, the aluminium mouldings must fold around the reinforcing internal mouldings. The bending radius of half the sheet thickness which is normal for steel body parts cannot yet be achieved reliably by most alloys in Group 3. One makes do with a so-called 'rope hem' of about twice the radius. Although such a rope hem has certain advantages for passive safety, it is frequently rejected by stylists. A reliable hemming process for these alloys therefore continues to represent a challenge for metallographers and tool makers.

Group 4 alloys

The group of alloys whose main alloying elements are copper (Cu) and magnesium (Mg) is also one of the precipitation-hardening alloy systems and is distinguished by a combination of high strength and good form-ability. The silicon content serves primarily to accelerate the precipitation process, in order to–as with the Group 3 AlMgSi alloys–achieve an increase in strength during paint stoving by means of a 'bake-hardening' effect. To achieve the precipitation-hardening effect, these alloys must also be supplied and processed in a cold precipitation-hardened state.

The alloys in this Group are stretcher-strain-free; the folding capacity is just as limited as that of the AlMgSi alloys in Group 3 (see Table 1.6).

In the unpainted state, AlCuMg alloys are prone to perforation corrosion due to the high copper content. The more recent generations of this type of alloy, 2038 and 2008, however, have considerably lower copper content.

Group 5 alloys
These new bodysheet alloys based on AlMgCu(Zn) were developed several years ago in Japan apparently with the aim of combining the good formability of high-alloy AlMg alloys with the freedom from stretcher strains of the precipitation-hardenable alloys (23, 24).

The precipitation-hardening effect during paint stoving seems to be considerably smaller than in the case of Groups 3 and 4 alloys, but apparently it is possible largely to compensate for the strength reduction.

Experiments with the engine bonnet of a production car, the Mazda RX-7, demonstrate the successful application of the '30-30' alloy in an outer skin component (23).

Aluminium sheet of all the alloys described above is generally given a surface finish which is produced by pairs of ground rolls. It has been found that this surface is prone to seizing during deep-drawing of body parts even at moderate blank holder pressures. In addition, the directional roughness of the so-called 'mill-finish' surface makes the coefficient of friction in the sheet plane variable, so the material flows irregularly in the tool (22, 25).

Different sheet surfaces have been developed with the aid of special roll preparations which have a non-directional, isotropic surface microstructure, as with steel sheet. A characteristic of these special surfaces is that the lubricant is located in individual 'lubricant pockets', and when the drawing edge of the tool is applied, a lubricant pressure cushion occurs which prevents the sheet bar surface seizing. This means that lower blank holder pressures can be used, so the flow of material in the tool can be controlled optimally.

Past results have shown that the forming behaviour of aluminium sheet can be considerably improved in this way, and the number of rejects can be reduced markedly. Such special surfaces bear the brand names ISOMATT, ISOMILL, ISOTEX, and LASERTEX.

Another approach is being tried currently with suitable pre-coatings on the strip material. This is intended to provide not only a uniform surface with favourable tribological characteristics, but also improved compatibility with adhesive bonding techniques and subsequent painting processes.

The corrosion of aluminium and anti-corrosion measures in vehicle construction

F. Ostermann

2.1 PRINCIPLES

2.1.1 Factors influencing corrosion resistance

The reaction of metallic materials to their environment is called corrosion if it causes measurable changes in the material (corrosion evidence), which may lead to impairment of the operation of the component (corrosion damage).

The environmental conditions for road vehicles may be considered to be particularly aggressive due to road salt and exhaust emissions. Knowledge about the corrosion performance of materials used in vehicle construction is therefore essential for the appropriate application of materials.

Numerous factors must be considered when assessing the corrosion resistance of metallic materials, components and assemblies.

(1) Corrosion conditions
    type of corrosive medium, for example chloride content, pH value;
    oxygen content of the corrosive medium;
    duration of influence, dry phases, temperature;
    movement of corrosive medium, erosion, cavitation.
(2) Factors related to the materials
    passive layer or oxide layer, resistance, conductivity;
    alloy components;
    microstructure and dispersion state;
    cold forming level;
    surface quality.

(3) Design and production conditions
      crevices;
      cavities, 'dishes' without drainage;
      contact with alien metals, mechanical connections;
      welded joints;
      mechanical processing;
      claddings, coatings.

The complex nature of the corrosion problem makes it appropriate to study the electrochemical principles first.

2.1.2 Electrochemical reactions

The corrosion mechanism is determined mainly by electrochemical processes. The corrosion process in metals requires contact between ionically conducting media (electrolyte) and the electron-conducting metal. At the phase boundary between metal and electrolyte, an exchange of electrons takes place between the co-reactants (Fig. 2.1).

If a metal atom M^{z+} is leached out of the metal lattice by the electrolyte, it leaves its valence electrons ze^- behind in the metal lattice. This metal-dissolving stage of the reaction (oxidation stage) is called an anodic partial reaction:

$$M_{lattice} \rightarrow M^z + ze^- \tag{1}$$

To re-establish electroneutrality between the metal and the electrolyte, the charge at one reactant must be released from the electrolyte. In

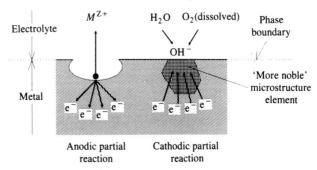

Fig. 2.1 Schematic representation of the electrochemical corrosion mechanism of metals in aerated, aqueous electrolytes

neutral electrolytes containing oxygen this reactant can be dissolved oxygen:

$$4e^- + O_2 + 2H_2O \rightarrow 4OH^- \tag{2}$$

This stage of the reaction (reduction stage) is called the cathodic partial reaction. The speed of the corrosion process is determined essentially by the amount of oxygen present.

The overall reactions in the corrosion process of iron or aluminium in a neutral, aerated, aqueous electrolyte are then as follows:

$$\text{iron:} \quad 2Fe + O_2 + 2H_2O \rightarrow 2Fe(OH)_2$$

$$\text{aluminium:} \quad 2Al + \tfrac{3}{2}O_2 + 3H_2O \rightarrow 2Al(OH)_3$$

It is true that the anodic and cathodic partial reactions take place at the metal/electrolyte phase boundary, but they do not necessarily have to take place at the same location on the surface. Anodes and cathodes may be distributed differently over the surface and thus produce different corrosion patterns:

uniform distribution   →   uniform surface removal

heterogeneous distribution   →   perforation corrosion,
selective corrosion.

Heterogeneous pit generation is associated with passive or protective surface layers.

2.1.3   Free and critical corrosion potentials (1)

When a metal is in contact with the corrosive medium (electrolyte), a constant potential value will occur in time, the so-called 'free corrosion potential' or 'rest potential' U_R. The value of the rest potential is dependent on the corrosion conditions mentioned above and on material-related factors.

The occurrence of perforation corrosion is characterized by a critical threshold potential under the relevant corrosion conditions ('breakdown potential' U_D). The position of the free corrosion potential can be displaced by contact with foreign metals or by an external potential impressed from outside ('mixed potential'). If the mixed potential is displaced in the anodic (positive) direction as far as the vicinity of the threshold potential for pitting U_D or beyond, perforation corrosion is to be expected. If the mixed potential moves in the cathodic (negative) direction, however, the result is a cathodic protection effect against perforation corrosion.

2.1.4   Curves of current density against potential

The electrochemical metal loss is proportional to the density of the corrosion current of the anodic (metal-dissolving) partial reaction (equation (1)). In potentiostatic measurements of the corrosion process, however, the current for the cathodic partial reaction ('oxygen reduction', equation (2)) is measured simultaneously. The results are plotted in so-called 'curves of total current density against potential' and indicate the corrosion performance of a metal in the electrolyte concerned (Fig. 2.2) (1, 2). Metals with a 'passive' response to a corrosive medium exhibit a distinct plateau extending from the pitting breakdown potential U_D in the cathodic direction. The metals are resistant to perforation corrosion in

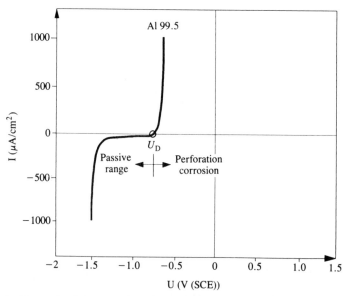

Fig. 2.2   Example of a curve of total current density against potential for a metal (Al99.5) with a distinct passive area in an aerated, aqueous electrolyte containing chloride. U_D = critical potential threshold for perforation corrosion

the area of this potential. On the anodic side of U_D the corrosion current, and thus the corrosion speed, increase exponentially.

2.1.5   Corrosive medium

A precise definition of the corrosive medium, in addition to numerous other factors, is crucial to the assessment of the corrosion resistance of materials. The difficulty of the assessment is also to be found here, with regard to the real behaviour of components, for example of a road vehicle. For this reason, acceleration of the real long-term effects of aggressive environmental conditions in short-term laboratory tests can give only approximate results. For the same reason, comparison of various materials with regard to their corrosion resistance is full of uncertainties.

The aggressive action of chlorides in aqueous corrosion media is undisputed, however. Chloride ions in the electrolyte weaken or penetrate the protective, passivating surface layer and put the metal in contact with the electrolyte, thus actuating the corrosion mechanism.

Artificial sea water is therefore used in corrosion studies as a standard corrosive medium for purposes of comparison.

2.1.6   The corrosion performance of different metals

Even if the basic electrochemical reactions of the corrosion mechanism are the same for all metals, the corrosion behaviour of different metals can vary in important points. Metals, such as aluminium or chrome/nickel steels, which have a tenacious, protective oxide layer or form a passive layer, are prone to pitting-type and selective corrosion in media containing chloride, such as sea water. In contrast, zinc exhibits surface corrosion in these media, although it develops a dense protective layer in the atmosphere. For this reason it is not necessarily possible to transfer successful anti-corrosion measures from one material to another.

Figure 2.3 shows the considerably lower corrosion resistance of non-alloyed steel in sea water when compared with various aluminium alloys. At the same time, the effect of different alloy compositions on corrosion resistance can be seen within the group of aluminium alloys. Furthermore, the somewhat parabolic course of the corrosion process with aluminium alloys points to a declining corrosion rate or increasing passivation of the surface (3).

Aluminium differs significantly from non-alloyed steels in corrosion performance in that a protective oxide layer is formed. The corrosion

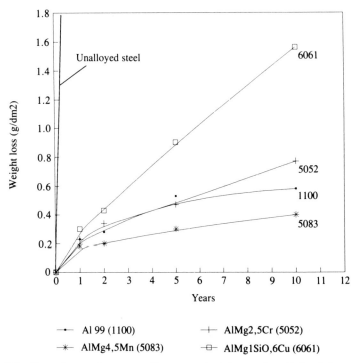

Fig. 2.3 **Corrosion performance of some engineering grades of aluminium and of non-alloyed steel in sea water. The ageing test took place on Harbor Island, NC, USA (3)**

performance of aluminium is therefore determined essentially by the effect of the various factors on the resistance of this oxide layer itself (see section 1.1).

2.2 THE OXIDE LAYER ON ALUMINIUM

2.2.1 Structure and significance

In air, metallically bright aluminium becomes coated spontaneously with a thin, dense and tenacious protective layer which has a very low capacity to conduct electrons and ions. The surface is thus passivated. This is the reason why aluminium is a remarkably corrosion-resistant metal, although it is a metal with high reaction energy (see electrochemical series of metals in section 4.4).

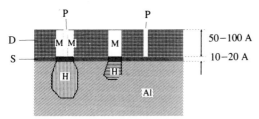

Fig. 2.4 **Schematic structure of oxide layer formed on aluminium in moist air at room temperature. Heterogeneous microstructure constituents, such as intermetallic phases, form mixed oxides. Al = base metal; D = top film; H = Heterogeneous microstructure constituents: M = mixed oxides; S = barrier layer; P = pores**

Figure 2.4 shows schematically the structure of an oxide layer formed in moist air at room temperature with a typical thickness of 0.005–0.01 μm. It consists of two parts, an almost pore-free barrier film (S) and a water-containing, porous top film (D).

The growth rate of the top film is promoted by humidity and increased temperature. It declines with time. The structure of the top film, which is amorphous at low temperatures, takes on an increasingly crystalline nature at higher temperatures, with a correspondingly greater chemical, electrical, and mechanical resistance. Top films formed as a result of hot forming and annealing treatment, therefore, frequently react differently from mechanically or chemically treated surfaces.

The oxides of alloying elements and microstructure constituents (for example precipitation phases) are incorporated in the oxide layer (mixed oxides) and can modify the mechanical, chemical, and electrochemical behaviour. This leads to the following conclusions which are important for the processing and application of aluminium.

(1) The high electrical resistance of the oxide layer is one of the causes of the shorter service life of spot-welding electrodes compared with steel. The surface resistance is dependent on the alloy composition and the manufacturing process of the wrought product.

(2) Chemical or mechanical surface pretreatment is necessary in order to achieve surface states specified for the adhesion of organic coatings and adhesives.

(3) Under critical ambient conditions, the corrosion resistance of aluminium is dependent on the alloy composition (see Fig. 2.3).

2.2.2   Chemical resistance

(a)   Natural repair of mechanical damage
Damage to the oxide layer due to scratches, stone impact, etc., does not
impair the chemical resistance, as a protective oxide layer again forms
spontaneously on the bright metal surface, as can be seen in Fig. 2.5.
Figure 2.5 shows the results of potentiostatic measurements of the
current density of a specimen of AlMg2Mn0.8 alloy in synthetic sea
water. When the surface is damaged, there is only a brief reaction of the
metal with the sea water before another protective oxide layer forms (**4**).

(b)   Solubility of the oxide film
The solubility of the oxide film in highly acid media (with the exception
of concentrated nitric acid and acetic acid) and in highly alkaline media
(except ammonium hydroxide) leads to surface metal loss. In these pH
ranges, aluminium can be used only to a very limited extent.

In the pH range between 4.5 and 8.5, the oxide layer is largely insolu-
ble (Fig. 2.6) (Shatalov; cited in Hatch (**5**)). Corrosion due to surface
metal loss is negligible. It is true that local corrosion, such as perforation
corrosion, can be caused by electrochemical processes in this passive
range, if the rest potential is anodically displaced. This can occur due to

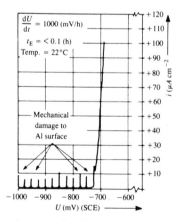

**Fig. 2.5   Diagram of total current density against potential for AlMg2Mn0.8 with
the potential-dependent passive and perforation corrosion range for synthetic sea water (4)**

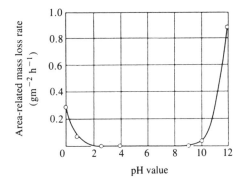

Fig. 2.6 Solubility of the aluminium oxide layer as a function of pH value (from Shatalov)

alloying elements dissolved in the mixed crystal which are more electropositive than aluminium, such as copper, for example (see potential series in section 4.4), due to heterogeneous microstructure constituents in the mixed crystal (Table 2.1, section 2.3.1), or to foreign metallic particles (grinding dust, brake dust) which have penetrated the surface.

Table 2.1 Electrochemical potential values of intermetallic phases and metallic precipitation elements in aluminium alloys (9)

Alloying element	Microstructure constituents	Potential, mV (SCE)*	
		pH 7.5	*pH 4.5*
Fe	Al_3Fe	−580	−660
Mn	Al_6Mn	−880	−800
Mn + Fe	$Al_6(Mn,Fe)$	−830	−800
Mg	Al_8Mg_5	−1380	−1090
Si	Si	−500	−580
Si + Fe	$Al_{12}Fe_3Si$	−720	−720
Mg + Si	Mg_2Si	−1230	−1240
Cu	Al_2Cu	−620	−620
	Cu	−230	−260

* SCE = saturated calomel electrode
Values at pH 4.5 in 3% NaCl solution, otherwise in artificial seawater at pH 7.5

2.3   THE EFFECT OF THE ALLOYING ELEMENTS ON CHEMICAL PERFORMANCE

2.3.1   General principles

The main alloying elements in aluminium alloys are:

$$Mg, Si, Mn, Zn, Cu, Fe, Cr$$

Depending on concentration and heat treatment state, they can occur both dissolved in the mixed crystal and in the form of heterogeneous microstructure constituents (precipitations, intermetallic phases). In the case of AlSi and AlSiMg casting alloys, silicon also occurs metallically as particles in the microstructure. The alloying elements have a varying effect on the chemical performance depending on the type, the quantity dissolved, the precipitated phase and distribution in the microstructure.

The effects of alloy admixtures on the following are important for influencing the chemical performance:

– the structure of the oxide layer, see Fig. 2.4;
– the extent of the passive range in the electrolyte concerned;
– the potential position (rest potential) in practical use of the aluminium material in the electrolyte;
– the potential difference between precipitation phases and the surrounding mixed crystal.

Table 2.1 lists the potential values for metallic and intermetallic heterogeneous microstructure constituents which may occur in aluminium alloys, depending on composition (**9**).

Intermetallic phases in the microstructure can be 'nobler' than the surrounding mixed crystal and thus shift the potential in the anodic, i.e., the positive, direction. As their proportion increases they thus promote the occurrence of perforation corrosion. However, they can also be 'less noble' or neutral with regard to the mixed crystal, and then only affect the corrosion performance if they lie on the metal surface and impair the resistance of the oxide layer. For most aluminium alloys, the potential of the mixed crystal in aerated synthetic sea water is between -1250 and -650 mV (SCE) (**10**).

Local corrosion such as, for example, perforation corrosion, in the pH range between pH 4.5 and pH 8.5 has its origin in the oxidation effect of the corrosive medium. Alloying elements or precipitation phases can

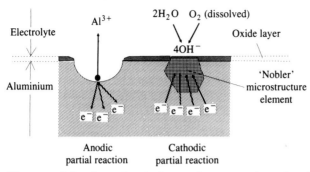

Fig. 2.7 Diagram of the electrochemical corrosion process in perforation corrosion of aluminium in aerated water

promote the oxidation reaction, because the air oxygen of the aqueous electrolyte is reduced to hydroxyl ions (HO^-) by them (Fig. 2.7).

2.3.2 Corrosion resistance of the most important alloys

The following types of corrosion can occur with aluminium and its alloys.

Pitting corrosion (PC) is electrolytic metal removal at isolated locations on the surface creates pits, of which the depth is usually greater than the diameter ('pitting').

Selective corrosion (SC) is a type of corrosion in which specific constituents of the microstructure, such as precipitations at grain boundaries, control the corrosion process.

Intercrystalline corrosion (IC) is a special form of selective corrosion which extends along the grain boundaries.

Exfoliation corrosion is a lamellar selective corrosion which can occur in the heat-affected zone of fusion welded bonds in wrought products of AlZn4.5Mg1-wa, if no additional heat treatment is performed after welding.

Stress corrosion cracking (SCC) is a type of corrosion in which a deformation-free, intercrystalline fracture occurs, often without visible corrosion, under the simultaneous and continuous influence of mechanical stress and aggressive media.

Table 2.2 gives a comparative assessment of the corrosion resistance of various aluminium alloys. The assessment of resistance in the table applies to unprotected components in a chloride-containing environment (salt water, road salt), and is based on general experience. Strictly speak-

Table 2.2 Corrosion resistance of selected aluminium alloys for vehicle construction

Alloy	Resistance to various types of corrosion		
	Pitting corrosion PC	Intercrystal corrosion IC(SC)	Stress corrosion cracking SCC
Al99.98	Excellent	n.s.	n.s.
Al99.5	Very good	n.s.	n.s.
AlMg2Mn	Very good	n.s.	n.s.
AlMg3	Very good	Good	n.s.
AlMg5Mn	Very good	Good	Good
AlMgSi0.5 wa	Very good	Good	n.s.
AlMgSi0.7 wa	Good	Good	n.s.
AlMgSi1 wa	Moderate	Moderate	n.s.
AlMgSiCu	Good	Moderate	n.s.
AlMg0.4Si1.2 ka	Moderate	Moderate	n.s.
GK-AiSi7Mg	Good	Moderate	n.s.
GK-AiSi10(Mg)	Moderate	Moderate	n.s.
G-AiSi12	Moderate	Moderate	n.s.
GD-AiSi8Cu3	Very mild	Mild	n.s.
AlZn4.5Mg1 wa	Moderate	Good	Moderate
AlCuMg1 ka	Mild	Mild	Moderate
AlCuMg2 ka	Very mild	Mild	Moderate

n.s. = not susceptible

ing, the assessment applies only to the group of aluminium materials themselves, and cannot be used for comparison with the behaviour of other metals.

The negative effect of Cu on the corrosion resistance of aluminium is due to the fact that the Cu potential is nobler than hydrogen. Thus, Cu which has gone into solution is deposited on the aluminium surface and forms a strong local cell. Aluminium alloys for vehicle construction which contain more than 0.3 per cent by weight of copper should therefore be given anti-corrosion treatment in applications which are susceptible to corrosion.

The merely 'moderate' resistance of silicon-containing cast alloys refers to the behaviour of worked surfaces of castings. The unprocessed casting skin has considerably greater resistance.

2.4 CORROSION PHENOMENA AND THEIR PREVENTION

Some types of corrosion which can play a part in aluminium materials for vehicles are described in more detail below, and ways of preventing them are indicated.

2.4.1 Pitting corrosion

For pitting corrosion of highest-grade and high-grade aluminium as well as Cu-free and Zn-free cast alloys and wrought alloys, the threshold potential is approximately the same. When wetted with chloride-containing water, for example sea water, it is between about -750 and -720 mV (SCE). This means that the aluminium matrix has approximately the same local solubility in such types of water. In practice, however, there can nevertheless be differences in corrosion performance, because, apart from the degree of purity, the precipitated intermetallic phases in the aluminium matrix or at the grain boundaries can exert an unfavourable influence.

Pitting corrosion can also be initiated by a heavy metal deposit on the surface or by iron particles (such as machining dust) embedded in the surface.

Pit growth declines with time, in accordance with an approximately parabolic time law, in otherwise unchanged corrosion conditions (l = pit depth, K = constant, t = time):

$$l = K \cdot t^{1/3}$$

In this connection see the curves in Fig. 2.3.

The pit bottom is anodic. With a high pit density the cathodic surface range declines; pit growth is decelerated by this. General experience shows that the greater the pit density, generally the smaller the pit depth.

Measurement of weight loss normally performed in corrosion tests does not give an adequate picture of the extent of the corrosion attack if there is pitting corrosion or selective corrosion. Such measurements must therefore be complemented by metallographic investigations, in which the maximum depth of attack, the proportion of the surface which is corroded, and the pit density are determined–see Figs 2.8 and 2.9 for an aluminium alloy and stainless steel after corrosion tests in drinking water containing chloride (6).

Prevention of pitting corrosion

– anodic oxidation (anodization);
– coating with organic polymers after suitable pretreatment;

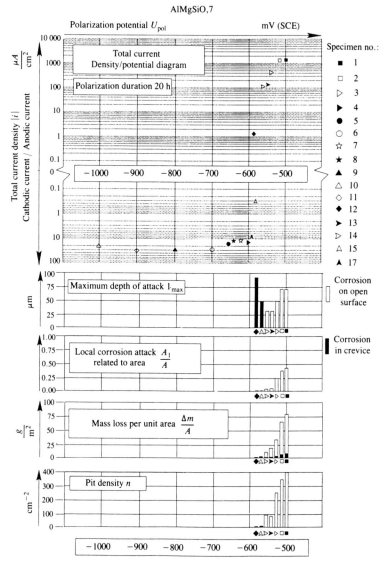

Fig. 2.8   Diagram of total current density against potential for the system AlMgSi0.7/drinking water (6). The bar charts below indicate corrosion values to DIN 50905

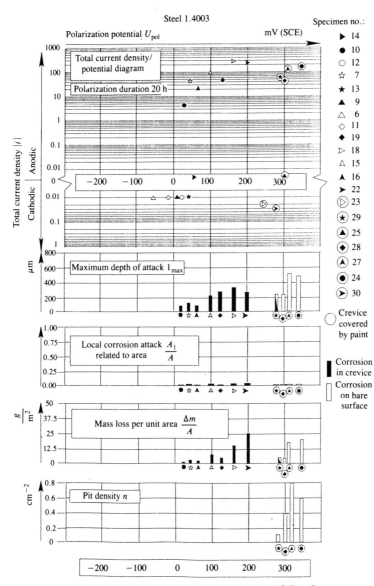

Fig. 2.9   Diagram of total current density against potential for the system stainless steel 1.4003/drinking water (6). When comparing the corrosion values with those in Fig. 2.8, note the different scales

- cathodic protection with galvanic anodes, such as zinc washers to compensate for the effect of stainless steel on aluminium.

2.4.2 Intercrystalline corrosion

Selective corrosive attack on precipitates at grain boundaries or in regions near grain boundaries, if the precipitates have a lower solution potential than the aluminium matrix.

Low-alloy materials, such as Al99.5, AlMn, AlMg (Mg < 3 per cent), and AlMgSi0.5 are largely insensitive to intercrystalline corrosion. Excess silicon in higher alloyed AlMgSi materials causes a certain insensitivity to intercrystalline corrosion.

Continuous precipitation seams at grain boundaries are particularly susceptible to intercrystalline corrosion. Such alloys are heterogenized by the manufacturer using appropriate heat treatments so that continuous seams are avoided (for example AlMg5Mn) and the precipitates at the grain boundaries resemble a string of beads.

Age-hardenable AlCuMg materials must be quenched as quickly as possible after solution annealing to prevent intercrystalline corrosion.

Prevention of intercrystalline corrosion
Apart from measures taken during production of semi-finished products, the same surface protection measures as are used for pitting corrosion also can be used to prevent intercrystalline corrosion (anodizing, coating, cathodic protection with galvanic anodes).

2.4.3 Crevice corrosion

Corrosion occurs in crevices existing for reasons of design or production, where there are critical crevice widths of about 0.02–0.5 mm.

Crevice corrosion occurs due to an inadequate supply of atmospheric oxygen (inadequate ventilation) in the crevice which is filled with the aqueous corrosive medium. The reduced oxygen concentration which exists within the crevice creates a corrosion cell between the outer surfaces which have greater ventilation and the inner anodic, crevice surfaces. This type of corrosion occurs on almost all metals, and also on metal/plastic combinations.

The critical threshold potential for crevice corrosion U_{Sp} takes precedence over the threshold potential for pitting corrosion U_D, that is the danger of crevice corrosion is greater than that of pitting corrosion, the greater the difference between the two threshold potentials. This applies especially to stainless steels, as measurements from a series of tests in chloride-containing drinking water show (**6**), in which the free corrosion potential U_R lies in the crevice corrosion range: stainless steel 1.4003 (Fig. 2.9) (**6**):

U_D = c. 0.3 V (SCE)
U_R = c. 0.075 V (SCE)
U_{Sp} = c. 0.03 V (SCE)

As the studies also show (**6**), due to the acidification of the electrolyte in the crevice in the stainless steel studied, the corrosion potential falls to about −0.36 V (SCE) after a long period, i.e. the steel changes from a passive state to an active one. Presumably repassivation of the steel is prevented by inadequate oxygen in the crevice (**6**).

In the case of aluminium the threshold potentials for crevice and pitting corrosion are close together: AlMgSi0.7 F27 (Figs 2.8 and 2.10) (**6**):

U_D = c. −0.57 to −0.56 V (SCE)
U_{Sp} = c. −0.59 to −0.58 V (SCE)
U_R = c. −0.59 V (SCE)

The free corrosion potential of the aluminium material studied falls in the course of time to values of about −0.7 V (SCE), i.e. into the passive range. Crevice corrosion attack declines accordingly (**6**), and with aluminium takes the form of pitting corrosion which is relatively shallow.

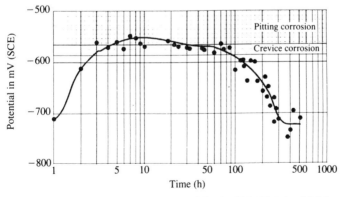

Fig. 2.10   **Curve of the free corrosion potential of AlMgSi0.7 in drinking water containing chloride measured in a crevice with plexiglass (6). Passivation of metal surface occurs after about 100 hours**

Prevention of crevice corrosion
In particularly critical cases for corrosion:

(1) crevices should be avoided in design and production;
(2) crevices should be filled or sealed off with permanently flexible sealants.

2.4.4 Bimetallic corrosion

Bimetallic corrosion can occur if two or more metals with different positions in the electrochemical series (see Table 2.3) are in metallic contact while simultaneously being wetted by an aqueous medium (the prerequisite being an electron conducting and ionically conducting contact) (1, 2).

Taking account of the values for the relevant pitting potential, the practical electrochemical series can give an indication of the risk of corrosion for combinations of metals; although this is only true to a limited extent. In practical conditions, the potentials or potential differences occurring depend on many factors:

– oxygen content of the electrolyte (contact between aluminium and 'nobler' metals, such as copper etc., are much less problematic if the electrolyte contains no oxygen);

Table 2.3   Electrochemical series of selected metals (7)

'Normal' potential*		Practical potential*			
		in water (pH 6)		in sea water (pH 7.5)	
Metal	mV	Metal	mV	Metal	mV
Copper	+95	Brass MS63	+100	Nickel	+1
Lead	−371	Copper	+95	Brass MS63	−32
Tin	−385	Nickel	+73	Copper	−35
Nickel	−475	V2A steel	(−129)	V2A steel	(−90)
Iron	−685	Aluminium	(−214)	Lead	−304
Zinc	−1008	Hard chromium plating	(−294)	Hard chromium plating	(−336)
Aluminium	−1905	Tin Sn 98	(−320)	Steel	(−380)
		Lead 99.9	(−328)	Aluminium	(−712)
		Steel	−395	Zinc	−851
		Zinc	−852	Tin	−854

* Measured against saturated calomel electrode
  Values in parentheses take account of a passivation layer

- composition, concentration, pH value of the electrolyte;
- aeration of the electrolyte and the electrodes;
- temperature.

In assessing the probability and the intensity of the occurrence of bimetallic corrosion, the following factors are important.

(1) Magnitude of the potential difference between the contact partners measured under appropriate test conditions (see electrochemical series). The anodic partner is at risk of corrosion.

(2) The magnitude of the electrical resistance between the contact partners. The greater the resistance, the lower is the risk of bimetallic corrosion.

(3) Presence of an electrolyte with appropriate aggressive reactivity and conductivity. The occurrence of condensation water should also be taken into account, as it can become an aggressive electrolyte when it collects dirt (road salt).

(4) Duration of the action of the electrolyte. Complete drying (ventilation, engine heat) facilitates renewal of the protective coating on aluminium.

(5) The area ratios of the contact partners determine, among other things, the current density (total current density of the anodic and cathodic partial reactions) of the electrochemical reaction. It is better for the 'nobler' contact partner to be smaller in area than the 'more base' partner.

To assess whether combinations of metals are favourable or unfavourable, one can also use the curves of current density against potential, see Fig. 2.2, for the relevant metals, so long as they have been measured in appropriate electrolytes. Figure 2.11 gives curves of current density against potential, measured in a 5 per cent NaCl solution, for several materials (2).

When two metals with different free potentials are combined, there is a mixed potential, in which the anodic net current of the 'more base' partner is equal to the cathodic net current of the 'less base' partner, to meet the condition of electroneutrality. Taking account of the area ratio of the contact partners, the magnitude of the anodic corrosion current and the position of the mixed potential can be calculated from the pairs

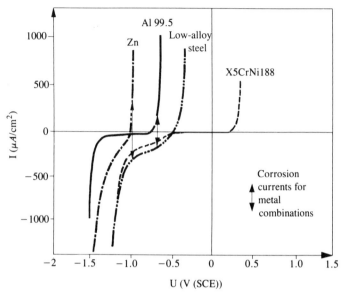

Fig. 2.11   Curves of total current density against potential, from Strobl (2), for aluminium, zinc and for non-alloyed and stainless steel, measured in a 5% NaCl solution against a saturated calomel electrode (SCE). Arrows indicate corrosion currents on contact between the relevant metals

of curves. Where the areas of the partners are equal, the values can be taken directly from the diagrams; see current arrows in Fig. 2.11.

Prevention of bimetallic corrosion
(1) Zinc generally protects aluminium cathodically and is attacked preferentially. Suitable contact partners for aluminium therefore are galvanized steel parts (sheet parts, bolts, nuts, washers etc.). The zinc layer must be sufficiently thick to ensure good long-term performance.
(2) Under normal conditions, periodic dry times leading to repassivation), contact between special steel parts (e.g. bolts) and aluminium is not risky. (Observe area ratios.)
(3) In favourable conditions (good ventilation, position on component, small area ratio), even contact between brass and aluminium is not a problem–for example, brass fittings on exposed aluminium petrol tanks and compressed air reservoirs on trucks.

(4) Carbon (graphite) attacks aluminium. Foam-rubber seals containing graphite should be avoided, even on painted parts. Damage to the paint coat, for example by stone impact, causes an unfavourable area ratio and thus leads to corrosion at the damaged point.

(5) Surface coatings (e.g. wax, oil, primer, paint) prevent or delay access of the electrolyte to the metal and hence also bimetallic corrosion.

(6) It is advisable for the 'nobler' contact partner to be isolated from the electrolyte by a non-conducting (organic) coating. Conversely, when the coating is damaged, more severe corrosion is to be expected (area ratio).

(7) In critical cases, the two contact partners should be electrically insulated from one another (plastic interface etc.). Care should be taken to avoid a shunt connection.

(8) Fasteners (bolts etc.) for joining different metals (such as steel to aluminium) should be of 'nobler' metal, preferably with an aluminous surface.

(9) At the design stage, avoid crevices, blind holes, 'troughs' etc. on the contact surfaces to prevent the collection of electrolyte. If necessary, provide drain holes and fill crevices with permanently flexible substances.

Surface treatment of aluminium components for vehicles

A. Blecher

Coating of aluminium and its alloys is performed for a variety of reasons. Those worthy of note are long-term corrosion resistance, decorative appearance and the achievement of new technological advantages while maintaining the outstanding physical and chemical characteristics of the light metal.

The quality of the coating of aluminium materials – be they rolled semi-finished products, extruded sections or castings – is determined decisively by the quality of the partly alloy-specific pre-treatment. Compromises or errors in pretreatment cannot be remedied by even the best coating system.

The main task of a mechanical or chemical pre-treatment is to produce a precision surface finish. Degreasing by wiping, immersion, and steam provide (in the order given) increasingly grease-free surfaces, without removing the oxide skin. Grinding, brushing, blasting, or polishing do not remove the oxide layer completely and generally serve as an initial stage for further surface treatment. Possible sources of defects for subsequent damage are 'rubbing in' of foreign metals into the aluminium surface, the use of brushes of brass or non-stainless steel, as well as sand and steel pellets. Chemical degreasing agents with a pickling effect or pickles remove the oxide layer and thus all impurities as well. After alkaline pickling of AlMg or AlSi alloys, the pickling deposits must be removed by acid post-treatment with nitric acid or nitric acid/ hydrofluoric acid (Figs 3.1 and 3.2 illustrate this for GD-AlSi12).

Despite careful degreasing by pickling, a painted aluminium surface can exhibit a lack of adhesion after a certain length of time due to environmental effects. Only a conversion coating, which forms by the reaction of chromium-containing or chromium-free chemicals with the metal, passivates the aluminium surface and protects it from the water vapour which diffuses through any paint layer. For aluminium surfaces,

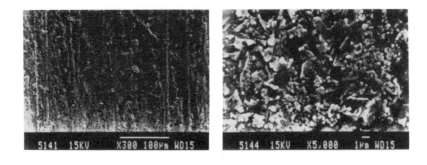

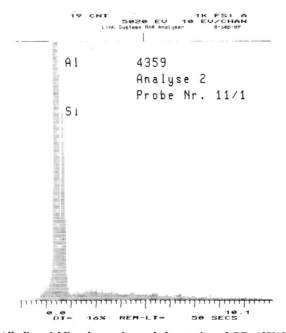

Fig. 3.1   Alkaline pickling degreasing and chromating of GD-AlSi12 *Top*: **surface in SEM at X300 and X5000 magnification.** *Bottom*: **EDX analysis**

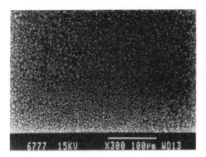
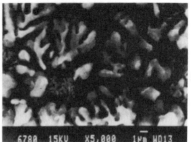

Fig. 3.2 **Alkaline pickling degreasing, dipping and chromating of GD-AlSi12** *Top*: **surface in SEM at X300 and X5000 magnification.** *Bottom*: **EDX analysis**

Table 3.1   Conversion treatment of aluminium

		Chromatizing		Phosphatizing on base	
Application		Yellow (Cr VI)	Green (No Cr V)	Zircon/ titanium	Zinc
Spray	'Rinse'	+	+	+	+
Dip	'Rinse'	+	+	+	(+)
Roll coat	'No rinse'	+	+	+	−

chromating, film-forming phosphatization, or a thin anodic oxide layer (2–5 μm) have proved successful with regard to providing adhesion and corrosion prevention. The various application methods for conversion coatings are given in Table 3.1.

In the case of the chromating methods, which can only be used for the separate treatment of aluminium parts, yellow and green chromating have been differentiated. Yellow chromating takes place at room temperature in a pre-treatment bath, which contains essentially chromic acid, active fluoride, and metal complexes such as ferricyanide (as an accelerator). The general composition of the yellowish iridescent to golden yellow coating can be described as follows: in the region close to the metal there are AlF_3, $AlOF$ and Al_2O_3; then follows (approximately 30 nm) $Cr_2O_3 \cdot XH_2O$ ($X = 1$ or 2), the top layer (2 nm) consists of $CrFe(CN)_6$. The yellow chromating layer also contains Cr^{VI} (see Figs 3.3 and 3.4).

To produce a green chromatization coating, which generally takes

1. Initial reactions

1.1. Pickling reaction
$$Al + 3H^+ \rightarrow Al^{3+} + 3H$$
$$Al^{3+} + 6F \rightarrow (AlF_6)^{3+}$$

1.2. Reduction of Cr^{6+}
$$(HCrO_4)^+ + 3H + 4H^+ \rightarrow Cr^{3+} + 4H_2O$$

2. Layer formation
$$Cr^{3+} + 2H_2O \rightarrow CrO(OH)\downarrow + 3H^+$$
$$Al^{3+} + 2H_2O \rightarrow AlO(OH)\downarrow + 3H^+$$
$$Cr^{3+} + [HCrO_4]^+ + 2H_2O \rightarrow Cr(OH)_2[HCrO_4]\downarrow + 2H^+$$
$$Cr^{3+}[Fe(CN)_6]^{3+} \rightarrow Cr[Fe(CN)_6]\downarrow$$

Fig. 3.3   Sequence of reactions in yellow chromating (schematic)

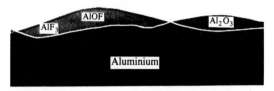

2 nm $\quad$ $Cr_2O_3 \times H_2O$ $\qquad$ $CrFe(CN)_6$

$Cr_2O_3 \times H_2O \times 1 \text{ or } 2$

30 nm

Fig. 3.4 Structure of a yellow chromatization layer, AlO (OH) 32.4%, H_2CrO_4 8.0%, $Cr(OH)_3$ 41.7%, $CrFe(CN)_6$ 17.9%

place at 40 to 50°C, chromic acid, active fluoride and phosphoric acid are used. The coat consists essentially of $CrPO_4 \cdot 4H_2O$ and demonstrably contains no Cr^{VI}. Depending on the weight of the coat, it is colourless to dull pale green. Today treatment times range from a few seconds to several minutes. Coat weight is determined by holding time and temperature, and primarily by the F^-/CrO_3 ratio (see Figs 3.5 and 3.6). Both chromium-containing conversion coatings produced by dipping or spray methods are optimal with regard to adhesion promotion and corrosion inhibition. Nevertheless, only a carefully prepared surface allows the formation of a homogeneous conversion coating. For example, pickling with silicate-inhibited pickling chemicals produces a thin silicate layer, which impedes formation of the chromatization coatings.

1. Initial reactions

1.1. Pickling reaction
$$Al + 3H^+ \rightarrow Al^{3+} + 3H$$
$$Al^{3+} + 6F \rightarrow (AlF_6)^{3+}$$

1.2. Reduction of Cr^{6+}
$$(HCrO_4)^+ + 3H + 4H^+ \rightarrow Cr^{3+} + 4H_2O$$

2. Layer formation
$$Cr^{3+} + (H_2PO_4) \rightarrow CrPO_4 \downarrow + 2H^+$$
$$Al^{3+} + (H_2PO_4) \rightarrow AlPO_4 \downarrow + 2H^+$$
$$AlPO_4 + Na^+ + F^+ \rightarrow Na\{AlPO_4(F)\}$$

Fig. 3.5 Sequence of reactions in green chromating (schematic)

$CrPO_4 \cdot 6H_2O$ 76–88%

$Al_2O_3 \ 3H_2O$ 12–23%

$CrPO_4 \cdot 4H_2O$

$Cr_2O_3 \times H_2O$ (≈10%)

in area close to metal

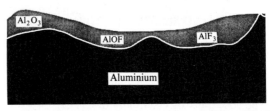

Fig. 3.6   Structure of a green chromatization coating, $CrPO_4 \cdot 6H_2O$ 76–88%, $AL_2O_3 \cdot 3H_2O$ 12–23%

Thorough rinsing is necessary after chromating. The final rinse should be performed with de-ionized water or the post-rinse solution recommended by the manufacturer. The sequence of events is summarized in Fig. 3.7.

In order to avoid the disposal of rinse water containing Cr^{VI} during the pre-treatment of aluminium strip, the 'no-rinse' methods which produce no waste water were developed. The chromium-containing pre-treatment solution, applied to the metal surface with an applicator roll

1. Cleaning and degreasing
2. Alkaline pickling
3. Rinse
4. Acid post-treatment HNO_3 or HNO_3/HF
5. Rinse
6. Conversion treatment
   (yellow or green chromating)
7. Rinse
8. Rinse with DI water
9. Dry in warm air

Subsequently prime
e.g. KTL coating

Fig. 3.7   Process sequence for chromating wrought Al materials

running in reverse, reacts completely, leaving no soluble salts; the aqueous residue is dried off at about 80°C. As no post-rinsing is necessary, the method produces no waste water. The coating weight can be influenced by altering the concentration of chemicals, roll velocity and roll pressure. Test results show that this method is equally as good as conventional chromating processes with regard to improvement of adhesion and corrosion inhibition. The 'no-rinse' methods are used primarily for pre-treating aluminium strip in practice.

In spite of these methods, the development of chromium-free pre-treatment processes is continuing in order to solve the environmental problems associated with the use of Cr^{VI} compounds. By using active fluorides and zirconium or titanium compounds or complex salts, chromium-free conversion coatings can be produced. Figure 3.8 shows the sequence of reactions. The layers consist of a mixture of aluminium oxide and an aluminium hydroxide, zirconium or titanium and fluoride complex. They are thin, colourless layers with acceptable adhesion and acceptable corrosion protection, which are produced at room temperature with dip, spray or roll application methods. The durability of these coatings for long-term corrosion protection still appears to require improvement.

Film-forming zinc-phosphating is particularly effective for conversion coatings in the treatment of aluminium semi-finished products, which are conductively bonded to steel or zinc parts; that is, of mixed construction.

In zinc-phosphating, phosphate layers form on iron and zinc, which contain iron, manganese or nickel ions depending on the base material or additives. In the initial pickling reaction (see Fig. 3.9) the base metal is dissolved. The consequent pH shift leads to the precipitation of zinc-phosphate which does not readily dissolve. When aluminium is passed through fluoride-free phosphating baths, no crystalline zinc-phosphate layers form. The aluminium surface is thoroughly cleaned by this

1. Pickling reaction
$$Al + 3H^+ \rightarrow Al^{3+} + 3H$$
$$Al^{3+} + 6F^+ \rightarrow (AlF_6)^{3+}$$

2. Layer formation
$$3(ZrF_6)^{2+} + 4(H_2PO_4) \rightarrow Zr_3(PO_4)_4 \downarrow + 18F^+ + 8H^+$$
$$(ZrF_6)^{2+} + 2H_2O \rightarrow ZrO_2 \downarrow + 6F^+ + 4H^+$$
$$ZrO_2 \times H_2O$$

Fig. 3.8 Sequence of reactions for zirconium-phosphating (schematic)

1. Pickling reaction

$$Al + 3H^+ \rightarrow Al^{3+} + 3H$$
$$Al^{3+} + 6F^- \rightarrow (AlF_6)^{3-}$$

2. Film formation

$$Zn^{2+} + H_2PO_4^- \rightarrow Zn(PO_4)^- + 2H^+$$
$$Zn^{2+} + 2Zn(PO_4)^- + 4H_2O \rightarrow Zn_3(PO_4)_2 + 4H_2O \downarrow \text{'Hopeite'}$$
$$M^{2+} + 2Zn(PO_4)^- \rightarrow Zn_2M(PO_4)_2 \downarrow M = Ni, Mn$$

3. Sludge formation

$$(AlF_6)^{3-} + 3Na^+ \rightarrow Na_3AlF_6 \text{ 'Cryolite'}$$

Fig. 3.9   Sequence of reactions in zinc-phosphating (schematic)

process, however. Film-forming zinc-phosphating of aluminium is only possible with phosphating solutions containing fluoride. The fluorides serve to complex the Al^3 ions which otherwise act as 'bath poison', and also act as a pickling accelerator. A crystalline zinc-phosphate layer, 1 to 3 g/m^2 in weight, now grows the same as on zinc. The pre-treatment can be by spray or immersion; zinc-phosphate coatings of various structures and morphologies are produced.

Spray pretreatment is preferred as some unwanted side effects can occur with immersion treatment which have so far proved unavoidable. The bath-operation system is relatively complex.

To increase corrosion resistance, a post-rinse is done with solutions containing Cr^{VI} or Cr^{III}. A further rinse with de-ionised water concludes the pre-treatment. The process sequence for zinc-phosphating is summarized in Fig. 3.10.

There are two methods of coating mixed constructions. One alternative is to fit the aluminium parts, then remove them from the body-in-

1. Alkaline cleaning
2. Weak alkaline cleaning with activation
3. Rinse
4. Zinc phosphating (film formation)
   (400−600 ppm F^-)
5. Rinse
6. Post-passivation (solutions containing Cr^{3+} or Cr^{6+})
7. Rinse with DI water
8. Dry with warm air

    Subsequently prime
    e.g. KTL coating

Fig. 3.10   Sequence of processes for zinc-phosphating Al components

white, give them separate pre-treatment and chromating, reassemble them with steel, phosphate the steel and aluminium parts with a fluoride-free zinc-phosphating process, KTL priming, filling and top-coat painting.

The second method consists of fluoride-free zinc-phosphating of both the steel and aluminium parts in the bodyshell, followed by the coating process described above.

The many methods of applying organic coatings do not differ from those for other metals and may be used for aluminium components without exception.

Roll application for coil coating and electropainting are the preserve of large-scale coating installations, while batch coating is performed essentially by electrostatic powder coating or by spraying with wet solvent-based paints.

It is true that aluminium has some special characteristics as a substrate with regard to electropainting. During anodic electrodipping of aluminium, secondary reactions at the anode produce a thin layer of aluminium oxide, which gives additional barrier characteristics to the fired paint layer.

Comparison of the corrosion-inhibiting characteristics of anodic and cathodic electropainting of aluminium after short-term tests indicates advantages for anodic paint deposition.

In comparable conditions for KTL deposition, coatings on aluminium as the substrate are basically thinner than on steel. The resultant coating thickness on aluminium is about 10 per cent thinner than those produced on steel, both for various thick-film KTL materials and for standard materials.

The fact that pre-treatment has taken up a large part of this chapter is a reflection of the importance of this preliminary stage for the subsequent coating process. A precision surface with a conversion coating produced according to the specification forms the ideal basis for all conventional coating methods.

Design perspectives for aluminium bodies

W. Wurl

4.1 OBJECTIVES

The desire for ever more comfortable vehicles with more special equipment, combined with higher engine power, has led in the past to a considerable increase in vehicle unladen weights. This trend has been promoted by desirable measures in the field of active and passive vehicle safety.

However, the simultaneous discussions and future conditions on fuel consumption should lead to a reduction in vehicle weight. Measures for reducing the weight of engine and chassis components, especially for expensive vehicles, have largely been exhausted now, so in future it will be necessary to give greater consideration to the still considerable potential for reducing the weight of the body.

4.2 REQUIREMENTS FOR THE VEHICLE BODY

As a load-bearing structure, the body must meet a great variety of requirements in all possible loading states.

During driving, a variety of forces are transmitted to the body, both quasi-statically and dynamically, via the chassis and the engine. Over the service life of the vehicle these must not lead to damage and should not cause vibration in the body which could impair safety or comfort. This means that the body must have high static and dynamic torsional and bending stiffness and rigid, durable force initiation points.

For crashworthiness, the passenger compartment must be strong and must have adequate energy absorption capacity in the front and rear structures and the doors.

The outer skin parts of the body should generally have a good surface finish combined with high buckling and shock resistance and non-aggressive crash performance, that is when damaged they should not

Table. 4.1   Requirements for the load-bearing structure

Area	Strength	Stiffness	Energy absorption
Floorpan	+ +	+ +	–
Crash area front and rear	+	+ +	+ +
Doors	+ +	+	+ +
Chassis/engine	+ +	+ +	–

splinter, as this could drastically increase the injury severity for persons involved in the accident.

Furthermore, there should be greater elastic deformation in the area of the bumpers to meet the various legal requirements. Tables 4.1 and 4.2 summarize the requirements.

4.3   DEFINITIONS

Strength
If one considers a supported flexible beam, as shown in Fig. 4.1, with a load F, the strength of the component is dependent on the maximum permissible stress σ_{zul} of the material, which must not be exceeded if it is not to break. The deformation f occurring during loading is of no significance for strength.

Stiffness
When assessing a component with regard to stiffness, it must be assumed that fundamentally the stresses occurring during loading lie below the

Table. 4.2   Requirements for outer skin parts

Area	Surface quality	Buckling resistance	Shock resistance	Elastic deformation
Horizontal components e.g., roof, bonnet	+ +	+ +	+	– –
Vertical components e.g., wings, doors	+ +	+	+	–
Functional parts e.g., bumpers	+	– –	+ +	+ +

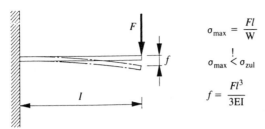

$$\sigma_{max} = \frac{Fl}{W}$$

$$\sigma_{max} \overset{!}{<} \sigma_{zul}$$

$$f = \frac{Fl^3}{3EI}$$

Fig. 4.1 Example of flexible beam. W = section modulus; I = second order moment of area; E = modulus of elasticity; σ_{zul} = allowable stress

yield point R_p and deformation thus occurs in the elastic range. The magnitude of the strength is found from the measured deformation f of the component, which, with the same geometry, is determined only by the modulus of elasticity of the material.

The stresses occurring during loading are of no significance for stiffness, so long as they are smaller than R_p.

Energy absorption capacity
The energy absorption is always dependent on the geometry of the component and the strength of the material. Specifying the geometry determines the deformation behaviour; for maximum energy absorption with minimum occupant loading, the deformation force F should be kept as constant as possible over the prescribed path x.

For the crush member shown in Fig. 4.2, the circular tube cross-section can be considered almost ideal with regard to crumpling.

The strength (yield point) of the material used determines, with the same deformation behaviour, the magnitude of the deformation force and thus the magnitude of the energy absorbing capacity of the component. In addition, the material should allow sufficient extension, as any other cracks or fractures occurring could cause the deformation force to fall suddenly or could have a negative effect on the deformation performance.

4.4 POSSIBLE SOLUTIONS

There are basically two ways of reducing the body mass while meeting the same requirements:

- optimizing the conventional sheet-steel design,
  or
- using lightweight materials.

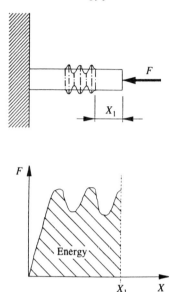

Fig. 4.2   Example of crush member

The conventional sheet-steel design for bodies has been largely optimized during recent years with the aid of electronic data processing methods, such as the finite element method, so there is little potential for future weight reduction here. The application of lightweight materials, as developed particularly in the aerospace industry, provides extensive possibilities for reducing the mass of vehicle bodies, as already demonstrated in various studies. Large-scale use has not been successfully implemented so far, mainly for reasons of cost. Only parts of the outer skin are made of aluminium or plastics on a medium to large scale by some manufacturers. Increased application of lightweight materials for frames and bodies is familiar from bus and truck construction.

In all cases, however, production methods and properties specific to the materials must be utilized in an optimum manner to keep investment and manufacturing costs to an acceptable level compared with those for steel sheet. The application of other materials while using the same parts geometry and production technology, such as, for example, replacing steel sheet with aluminium sheet, generally leads to considerably higher costs and unsatisfactory results.

4.5 MATERIALS CHARACTERISTICS

Figure 4.3 contains various materials and their characteristics compared with deep-drawing sheet steel St14. The various material-related features of pre-treatment and processing also lead to advantages or disadvantages for the individual material, depending on component geometry.

As a known lightweight material, aluminium has a very large mass-specific energy absorption capacity combined with good tensile strength. As it is a metallic material, similar to steel sheet, similar tools and equipment can be used for production and repair, which is usually an advantage when converting from one material to another.

The materials characteristics of GFRP are heavily dependent on the proportion and orientation of the reinforcing fibres, so the achievable

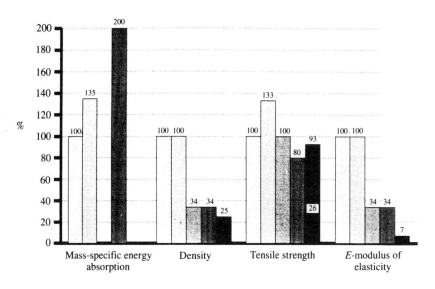

☐ ST14
☐ ZST E 260 (hot-rolled
   fine-grained sheet steel)
▨ AlMg4.5Mn F27
■ AlMgSiO.5 F22
■ GFRP (25/60% glass content)

Fig. 4.3   Characteristics of materials

strength must always be seen in conjunction with the complexity of the production process. It is scarcely possible to achieve the strength of St14 with the most economical processes known today.

One general disadvantage of lightweight materials is the low modulus of elasticity, which has a negative effect on the stiffness of the component.

4.6 DIMENSIONING WHILE MAINTAINING COMPONENT STRENGTH

For components which must be designed primarily for strength, initial dimensioning gives the following comparable material thicknesses compared with familiar steel sheet:

St14   1.00 mm
ZStE 260   0.75 mm
AlMg4.5Mn Γ27   1.00 mm
AlMgSi0.5 F22   1.25 mm
GFRP (25/60)   1.08 to 3.85 mm

This does not take account of possible differences in geometry dictated by the materials and the endurance limit of the component determined by the joining technology. Figure 4.4 gives a summary of an initial assessment of possible reductions in component mass.

The materials ZStE and GFRP allow mass reduction only to a limited extent; the production-related requirements of GFRP in particular should be borne in mind, however.

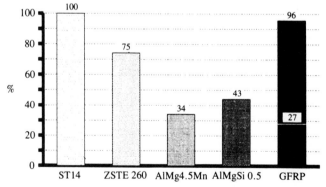

Fig. 4.4   Mass with equal strength

By using aluminium, the mass of components where strength is critical can be reduced by approximately 60 per cent. A further reduction is possible by using higher-strength alloys, though they must be assessed individually with regard to adequate weldability and corrosion resistance.

4.7 DIMENSIONING OF CRUSH MEMBERS

Aluminium offers outstanding opportunities in the design of crush members. Its extrudability facilitates the production of sections which are not possible with any other manufacturing process. The main advantage is that it is not necessary to join several parts in the longitudinal direction, as is necessary, for example, with members made of two sheet shells. It is thus impossible for there to be any uncontrolled opening along the seam.

Furthermore, the whole cross-section can be divided up into almost rectangular or even round-section parts by fitting intermediate bridges (Fig. 4.5), which, with a high and uniform level of force, effect optimal crumpling of the member after axial loading.

Tests have shown that an optimal section profile achieves higher energy absorption with a more uniform level of force with the same external dimensions than familiar steel sheet members with the same thickness of material. It is possible to achieve a reduction in mass of up to 75 per cent.

4.8 DIMENSIONING WITH THE SAME STIFFNESS

4.8.1 Torsional rigidity
When designing a load-bearing structure for a vehicle, in addition to crash safety, there must be adequate torsional rigidity. With open vehicles there are particular difficulties because the stiffening effect of the roof structure is absent. Experience has shown that good torsional rigidity almost automatically ensures adequate bending stiffness.

As high torsional rigidity with low mass can only be achieved with closed-section members, open sections are not considered below. An example is given of a square, thin-walled section with constant material thickness, as in Fig. 4.6.

$$c = \frac{E \times 2 \times A_m^2 \times s}{(1 + \mu) \times l \times U_m} \text{ torsional rigidity} \tag{1}$$

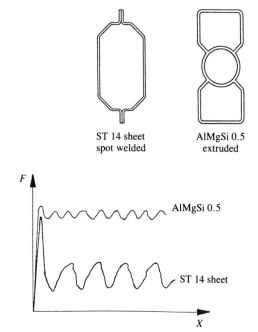

Fig. 4.5   Energy absorption of crush members

The 'simplified' calculation of the torsional rigidity of the section, as in equation (1), shows that both the modulus of elasticity and the material thickness are linearly proportional to the torsional rigidity, i.e., if the

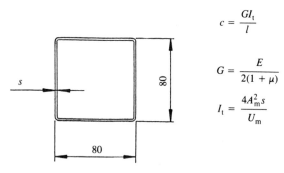

$$c = \frac{GI_t}{l}$$

$$G = \frac{E}{2(1 + \mu)}$$

$$I_t = \frac{4A_m^2 s}{U_m}$$

Fig. 4.6   Example of torsion member, section

modulus of elasticity or the material thickness is doubled, the torsional rigidity is doubled, assuming the same Poisson's ratio.

$$m = U_m \times s \times l \times \rho$$

$$\frac{c}{m} = \frac{E \times 2 \times A_m^2}{(1 + \mu) \times l^2 \times \rho \times U_m^2} \text{ mass-related torsional rigidity} \tag{2}$$

c	spring constant (torsional rigidity)
m	mass
ρ	density: steel 7.86 g/cm³; Al 2.70 g/cm³
l	length
G	modulus of elasticity in shear
E	modulus of elasticity: steel 210 000 N/mm²; Al 70 000 N/mm²
μ	Poisson's ratio: steel 0.3; Al 0.3
I_t	torsional moment of area
A_m	area bounded by the centreline
s	material thickness
U_m	length of centreline

A comparison of the steel and aluminium sections with regard to mass-specific torsional rigidity, equation (2), shows clearly that no reduction of mass is possible, as the modulus of elasticity and the density of aluminium are lower by about a factor of 3 than those of steel. However, this method of calculation disregards various criteria.

The material thickness of the aluminium member must in any case be greater than would be necessary for equal buckling rigidity to the steel member; see 4.8.2. In practical experiment, the higher buckling stiffness of the surfaces has a not inconsiderable influence on the torsional rigidity of a member, which is not taken into account in the 'simplified' calculation.

Aluminium makes it possible to manufacture sectional members by extrusion, and thus to dispense with joining. Members made of sheet shells (Fig. 4.7) are generally joined with spot welds at intervals of 40 to 80 mm, and are thus considerably more flexible in the region of the seam than 'perfectly joined' sections of extruded aluminium.

A further advantage of aluminium is that an extruded section needs no space for the spot-weld flange, and can thus have a larger closed cross-section with the same external dimensions. As the size of the cross-section has a quadratic influence on the torsional rigidity (see equations (1) and (2)), using extruded aluminium sections provides further potential

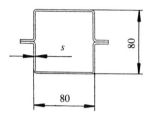

Fig. 4.7   Example of torsional member, sheet

for reducing weight. This is not considered in the remainder of the analysis.

The following material thicknesses were obtained for members with equal torsional rigidity in tests with similar sections.

St14   spot welded $s = 1.0$ mm
ZStE   spot welded $s = 1.0$ mm
AlMg4.5Mn   spot welded $s \approx 2.3$ mm
AlMgSi0.5   extruded $s \approx 2.0$ mm

For a GFRP member with no allowance for a joining seam, wall thicknesses of between 5 and 9 mm are necessary, depending on the quantity and orientation of the reinforcing fibres.

If one takes advantage of the material-related production advantages, it is possible to achieve a weight saving of approximately 42 per cent for aluminium with twice the thickness of material as with steel sheet (Fig. 4.8).

4.8.2 Buckling stiffness
In addition to influencing the torsional rigidity of section members (see 4.8.1), buckling stiffness is very important for outer skin parts. In particular, horizontal surfaces such as the bonnet and boot lid and the roof should be made of a material with high buckling stiffness, because there are few opportunities here for geometrical measures, such as beads. Often, the interior sheet of bonnets and boot lids has to be provided with large amounts of support, to prevent buckling of the lid when it is closed, with consequent weight disadvantages.

If one considers simply the buckling load on a sheet as a bending process of a beam supported at two points shown in Fig. 4.9, the necessary material thickness of various materials compared with steel sheet

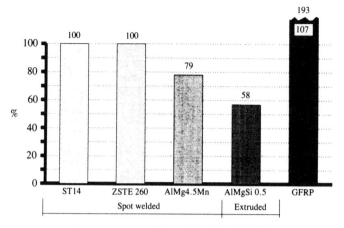

Fig. 4.8   Mass with equal torsional rigidity

can be calculated.

$$f = \frac{F \times l^3}{48 \times E \times I} \; ; \quad I = \frac{b \times s^3}{12}$$

$f_{St} = f_{Al}$ equal buckling stiffness steel – aluminium

$$\frac{F \times l^3}{48 \times E_{st} \times I_{st}} = \frac{F \times l^3}{48 \times E_{Al} \times I_{Al}}$$

$$S_{Al} = \sqrt[3]{\{(E_{St}/E_{Al}) \times S_{St}^3\}} \tag{3}$$

$$S_{Al} = 1.44 \times S_{St} \tag{4}$$

$$S_{GFRP} = 2.41 \times S_{St} \tag{5}$$

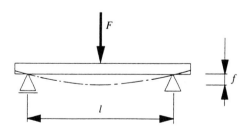

Fig. 4.9   Example, buckling stiffness

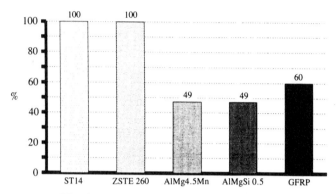

Fig. 4.10   Mass with equal buckling stiffness

Equations (4) and (5) give the weight reductions shown in Fig. 4.10 compared with steel sheet.

In the case of components which must be dimensioned for buckling stiffness, weight savings of about 51 per cent can be achieved with aluminium.

Before deciding to use GFRP, the production requirements and the difficulties involved in achieving a sufficiently good surface finish should be given especial consideration.

4.9   CONCLUSION

The load-bearing structure and the external skin parts of a body offer considerable potential for reducing vehicle weight. By using aluminium, all requirements for the body-in-white can be met with the exception of elastic deformability in the region of the bumpers.

The necessary dimensioning for aluminium is given for comparison with today's conventional steel-sheet design, to give the designer a point of reference to the current status.

Strength-relevant components are generally not critical, because it is necessary to increase the thickness of the material by only about 25 per cent compared with steel sheet. Greater demands can be met by using higher-strength aluminium alloys. Components which must be designed for torsional rigidity, such as the whole member structure of the vehicle, are already over-dimensioned with regard to strength in most areas when using low-strength, easily extrudable alloys.

The use of closed sections for torsionally rigid members of extruded aluminium offers a weight saving of about 42 per cent, despite the relatively low modulus of elasticity. Greater savings can be achieved by improved space utilization by enlarging the cross-sections.

The optimum design opportunities of an extruded crush member make it possible considerably to improve the function of the component compared with a steel-sheet design having minimal weight. The weight of external skin parts, which must have particularly high buckling stiffness, can be reduced by about 51 per cent with a 44 per cent increase in material thickness by using aluminium.

Various studies show that an aluminium body can meet all necessary requirements at about 50 per cent of the weight of a steel body. But the most important task is to devise an economical overall concept comparable with steel sheet by optimal utilization of aluminium's specific production advantages.

5

Joining aluminium body materials

B. Leuschen

5.1 INTRODUCTION

Weight reduction in vehicle construction can be achieved by reducing wall thicknesses and cross-sections by using stronger materials, and also by using materials with a lower specific weight, such as plastics and aluminium alloys. The main target for reducing vehicle weight is the body.

The methods given in Table 5.1 can be used for joining aluminium materials, and are described in more detail below, with the exception of adhesive bonding. Due to its high productivity, resistance spot welding (called briefly spot welding below), is the most important for mass production.

There are some essential differences between joining aluminium and steel which result from the physical characteristics of the two materials. The following characteristics are of particular interest for aluminium welding technology:

– low melting temperature;
– high affinity with oxygen (oxide film formation);
– good electrical conductivity;
– good thermal conductivity;
– high coefficient of thermal expansion.

New aluminium body sheet was developed at the end of the 1970s (1) as the aluminium materials known up to then did not meet one or more requirements of vehicle construction, such as adequate strength, high formability, satisfactory weldability, good painting properties, or favourable corrosion performance. The alloy AlMg0.4Si1.2 is used for exterior skin parts such as the bonnet, boot lid, doors and wings due to its freedom from stretcher lines, while AlMg5Mn, a material with good strength characteristics, is used for concealed interior parts such as the seat-cushion frame and reinforcements.

Table 5.1   Methods of joining aluminium components in body— construction

Joining process	Suitable for aluminium	Compared with joining sheet steel	
		Special pretreatment necessary*	Special equipment necessary
Resistance spot welding	Yes	Yes	Yes
Resistance projection welding	Yes	Yes	Yes
TIG welding	Yes	No	Yes
MIG welding	Yes	No	Yes
Laser welding	Limited	No	No
Adhesive bonding	Yes	Yes	No
Penetration bonding	Yes	No	No

* basically, the parts to be joined should be degreased.

Table 5.2   Aluminium body materials

	AlMg5Mn	AlMg0.4Si1.2
Sheet thickness (mm)	1.25 1.50 2.00	1.15 1.25 2.00
Type of alloy	Not age-hardenable	Age-hardenable
State	Soft	Cold age-hardened
Surface	Mill finish	Mill finish
Electrical conductivity (S · m/mm^2)	18	27
Thermal conductivity (W/m · K)	130	170
R_m (N/mm^2)	270	250
$R_{p0.2}$ (N/mm^2)	130	140
Stretcher lines	Yes	No
Application	Interior parts	External skin

The two materials differ not only in their state, but also primarily in their specific electrical conductivity and in their thermal conductivity (Table 5.2). While the soft AlMg5Mn material is one of the poorest conductors among aluminium alloys, the cold age-hardened AlMg0.4Si1.2 material is one of the best.

5.2   RESISTANCE SPOT WELDING

5.2.1   Spot-weldability

In spot welding, the components to be joined are pressed together locally by two copper electrodes and heated at the contact point by Joule resistance heating up to the welding temperature. The size of the welding point produced depends primarily on the thermal balance variation at the welding location during the welding process.

According to Joule's law, the welding current, the total resistance and the welding time determine the heat supply. Heat is removed primarily by thermal conduction into the water-cooled electrodes, into the sheet, and to a lesser extent, by thermal radiation. The heat loss depends, among other things, on the sheet thickness, the electrode shape, and especially on the specific thermal conductivity of the materials being joined. In the case of materials with high electrical and thermal conductivity, such as aluminium and its alloys, the variation of the energy supply over time also plays a part. This is determined predominantly by the type of welding current used (direct current, alternating current).

The total resistance consists of the contact resistances and the material resistances of both the electrodes and the parts being joined (Fig. 5.1). During current flow, heat is produced in accordance with Joule's law at each resistance in the current circuit, and it is produced in proportion to the ratio of the individual resistance to the total resistance. While the material resistances of the electrodes are negligibly small due to their good conductivity with regard to the other resistances independently of the material being welded, there are fundamental differences in the magnitudes of the contact resistance and material resistance of the parts being joined when using different materials for joining.

So, for example, aluminium materials exhibit distinctly lower material resistances than steel, and considerably higher welding currents are necessary to produce a spot-weld bond with aluminium. The electrical conductivity fluctuates depending on the type, proportion and precipitation state of the individual alloying elements. Steel and aluminium also

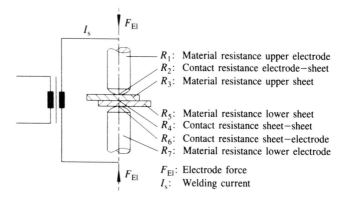

	R_1:	Material resistance upper electrode
	R_2:	Contact resistance electrode–sheet
	R_3:	Material resistance upper sheet
	R_5:	Material resistance lower sheet
	R_4:	Contact resistance sheet–sheet
	R_6:	Contact resistance sheet–electrode
	R_7:	Material resistance lower electrode
	F_{El}:	Electrode force
	I_s:	Welding current

Resistances before start of welding	Steel	Aluminium
R_1, R_7	negligibly small	
R_2, R_4, R_6	large	small
R_3, R_5	large	small

Fig. 5.1   Resistances involved in the spot welding of steel and aluminium

differ in the magnitude of the contact resistances. Aluminium alloys have high contact resistances due to the presence of an oxide film of different composition and thickness on each surface which is not electrically conductive. This oxide film breaks up during the welding process as a result of the surface deformation caused by the electrode force. The resulting contact bridges grow very rapidly with current flow and thus leads to a decline in contact resistances.

The spot-weldability of aluminium materials is determined mainly by the chemical composition, the metallurgical characteristics and the surface finish of the parts being welded (Fig. 5.2).

The chemical composition of the aluminium materials affects the electrical and thermal conductivity and hence also the parameters necessary for the welding process. As conductivity increases, the thermal loading on the electrode working surface increases and there is thus greater elec-

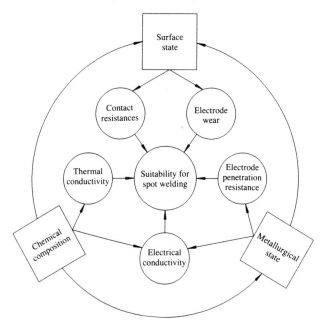

Fig. 5.2 Suitability of aluminium materials for spot welding

trode wear. Conductivity falls with an increasing content of alloying constituents.

The surface state (oxide film) of the parts to be joined has a great effect on the magnitude and continuity of the contact resistances and thus on the suitability for spot welding. The oxide films cause considerable heat production at the electrode/sheet contact points, as well as increased alloying of aluminium at the electrodes and, consequently, severe electrode wear. Nevertheless, it is possible to achieve sufficiently strong welded joints even with untreated aluminium sheet. To achieve uniform strength values and as large an electrode life duration as possible, however, the oxide film should be removed by surface treatment before welding. The contact resistance is affected by the oxide film, its thickness, composition and uniformity, by the roughness profile of the surface, and also by possible metallic coatings or impurities (grease, oil, dirt, dust).

The metallurgical condition of the parts to be joined (soft, strain-hardened, precipitation-hardened) directly affects their characteristics. With increasing strain-hardening, electrical conductivity falls slightly, but penetration resistance to the electrodes increases. Strain-hardening therefore improves suitability for spot welding.

5.2.2 Surface pre-treatment

Chemical methods (for example pickling) and mechanical methods (such as brushing) can be used to remove or reduce the oxide films.

The procedure for mechanical and chemical pretreatment is described in (**2**). Mechanical pre-treatment should be performed shortly before welding in as mechanized a manner as possible using rotating, non-rusting wire brushes. Chemical pretreatment consists of several stages, for example:

- degrease with mild alkaline agents;
- rinse (possibly several times);
- pickle (phosphoric acid);
- rinse (several times);
- dry (hot air).

The success of the pickling treatment depends decisively on matching the pickling agent to the initial state of the surface of the workpiece, which is determined essentially by alloy composition and manufacturing process.

The result of the surface treatment can be monitored by measuring the transition resistance. The procedure is described in (**3**). Figure 5.3 gives the results of resistance measurements on Al sheets subjected to different treatments. In the untreated state the sheets have high transition resistances, with a large degree of scatter. With surface treatment, the transition resistance can be reduced considerably in magnitude and also in the scatter of the values. To achieve reproducible welds with regard to strength and point formation, the transition resistances should be below 50 $\mu\Omega$ with this measuring method.

5.2.3 Parameter settings

The welding parameter settings will vary depending on the characteristics of the welding machine, the type of current, the electrode shape, the aluminium materials, the thickness of the parts to be joined, and the preparation of the parts. The approximate values listed in Table 5.3 serve as a starting point for determining suitable welding parameters. They apply to aluminium materials with low electrical conductivity, such as AlMg5Mn. It can be seen that, compared with steel, considerably higher

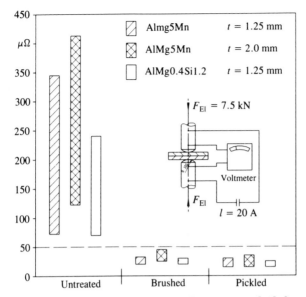

Fig. 5.3   Transition resistances for various pretreated Al-sheets

welding currents and shorter welding times are necessary. This is due to the good electrical and thermal conductivity. If alloys with higher electrical conductivity are used, the welding current must be increased. The current must be increased likewise if alternating current is used instead of direct current, while untreated sheets require lower currents than pretreated sheets due to the higher contact resistances in the sheet/sheet plane.

Swaged electrode caps (diameter ≥ 16 mm) made of CuCrZr or CuAg have proved suitable. Due to their greater hardness, CuCrZr electrodes have the advantage of slower deformation of the electrode working surface. CuAg electrodes have advantages where higher surface quality is required.

When selecting the electrode working surface, the requirements for the bonds must be taken into account. If alloying tendency and/or electrode penetration depth are the main criteria, it is advisable to use ball electrodes with a larger ball radius than for steel welding. If the main criterion is the continuing strength of the bond over as large as possible a number of spot welds, conical electrodes can also be used. Their disadvantage is increased electrode penetration into the surface of the sheet.

Table 5.3   Approximate values for spot welding with direct current for an alloy such as AlMg5Mn (4)

Settings	Sheet thickness (mm)											
	0.35	0.5	0.8	1.0	1.25	1.5	2.0	2.5	3.0	3.5
Electrode diameter (mm)	16	16	16	16	16	20	20	20	25	25
Electrode ball radius (mm)	75	75	75	75	100	100	100	100	100	150
Electrode force (kN)	1.5	1.8	2.2	3.0	3.5	4.0	5.0	6.5	8.0	10.0
Welding current (kA)	18–22	19–24	24–30	25–32	26–34	27–35	30–38	34–42	38–45	44–50
Welding time	2	2	3	3	4	5	6-8	7-9	8-10	9-12
Point diameter (mm)	3.0	3.5	4.5	5.0	5.5	6.0	7.0	8.0	8.5	9.5
Spot diameter (mm)	3.5	4.0	5.0	5.5	6.0	6.5	7.7	8.7	3.3	10.3

5.2.4 Point and spot diameter

As with spot welding of steel, one has to distinguish between point diameter and spot diameter with spot welding of aluminium. The point diameter is the diameter of molten material measured on the metallographic section. With untreated sheet, determining the point diameter is problematic, as the points are not usually circular. Spot diameter is measured after destructive testing. In a shear fracture, the spot diameter is the average diameter of the fracture area without adhesion zone. In a button fracture, the spot diameter is the average diameter of the basic area of the button-like bleb. Experience shows that the spot diameter is approximately 10 per cent larger than the point diameter. Unlike steel, with most Al-alloys the type of fracture is not a quality criterion. The strength of the welded joint depends primarily on the point or spot diameter. Table 5.4 gives the minimum necessary point and spot diameters for various sheet thicknesses. During manufacture, the point diameter reduces due to electrode wear. For this reason it is necessary to select welding parameters which give a larger point diameter with new electrodes.

5.2.5 Spot interval, edge interval, and overlap

Due to the good electrical conductivity of aluminium materials, the shunt effect is much more noticeable than with steel in spot welding of multi-spot specimens or components. The shunt current flows not only through the last-welded spot, but, depending on spot interval, through a larger number of already welded spots. Only when the spot interval reaches about 30 times the sheet thickness (AlMg5Mn) or 40 times the sheet thickness (AlMg0.4Si1.2) does the shunt cease to have a serious effect on spot diameter and thus on the strength of the welded joint. With smaller spot intervals it is necessary to counteract the shunt effect from the second spot weld by increasing the welding current.

To prevent liquid material spraying out of the weld point, it is necessary to have a sufficiently large edge interval and a correspondingly large overlap of the two sheets being joined (Table 5.4).

5.2.6 Load-bearing capacity

The spot-weld diameter or the tensile shear force determined in the free tensile shear test is generally used as a measure of the static load-bearing capacity of aluminium welded joints (5). The shear tension strength of aluminium spot-welded joints increases with increasing thickness of the parts being bonded and increasing tensile strength of the base material.

Table 5.4 Minimum values for point and spot diameter, edge interval, flange width, and overlap width

Material	AlMg5Mn							AlMg0.4Si1.2						
Sheet thickness, T	1.0	1.15	1.25	1.5	2.0	2.5	3.0	1.0	1.15	1.25	1.5	2.0	2.5	3.0
Minimum point diameter, d_l	4.0	4.3	4.5	4.9	5.7	6.3	6.9	4.5	4.8	5.0	5.5	6.3	7.1	7.8
Minimum spot diameter, d_p	4.4	4.7	5.0	5.4	6.3	7.0	7.6	5.0	5.3	5.5	6.0	7.0	7.8	8.6
Minimum edge spacing, v	5.5	5.9	6.3	6.8	7.9	8.8	9.5	6.3	6.6	6.9	7.5	8.8	9.8	10.8
Electrode diameter, d_E	←— 16 —→				←— 20 —→			←— 16 —→				←— 20 —→		
Minimum flange width	17.5	18.0	19.0	19.5	23.5	25.0	26.5	18.5	19.0	19.5	20.0	24.5	26.0	27.5
Electrode diameter, d_E	←— 16 —→				←— 20 —→			←— 16 —→				←— 20 —→		
Minimum overlap	13.0	14.0	14.5	15.5	18.0	19.5	21.0	14.5	15.5	16.0	17.0	19.5	21.5	23.5
(All values in mm)			$d_l \geq 4.0 \times \sqrt{t}$							$d_l \geq 4.5 \times \sqrt{t}$				

$$d_p = 1.1 \times d_l \qquad v \geq 1.25 \times d_p \qquad b \geq \frac{d_E}{2} + r + v + x \qquad r = 1.5 \times t \qquad \bar{u} = 2 \times v + y$$

x = flange width tolerance + flange offset + electrode offset = 2.5

y = sheet offset + electrode offset = 2.5

In the case of cold-worked and age-hardenable alloys, however, it must be remembered that the increase in strength achieved by cold-working or age-hardening is lost in the heat-affected zone due to the welding process. The shear tension forces of age-hardenable alloys can be increased by subsequent age-hardening.

Figure 5.4 shows the correlation between spot diameter and shear tension force. In the spot diameter range investigated, the shear tension force for all sheet thicknesses increases almost linearly with increasing spot diameter. With thicker sheet it is possible to achieve spot welds with a larger spot diameter, which produce correspondingly larger shear tension forces. The age-hardenable material AlMg0.4Si1.2 requires larger spot diameters than the AlMg5Mn material with the same sheet thickness for the same shear tension forces, as a result of the lower strength of the base material and the additional loss of strength due to the welding process.

Dynamic studies of one-spot specimens (Fig. 5.5) and of single-row lap multi-spot specimens show that the fatigue limit of Al alloys is reached at a minimum number of oscillation cycles of 2×10^7. While one-spot

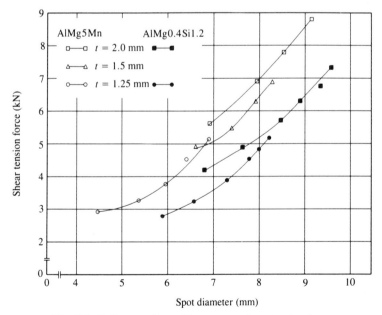

Fig. 5.4 Influence of spot diameter on shear tension force

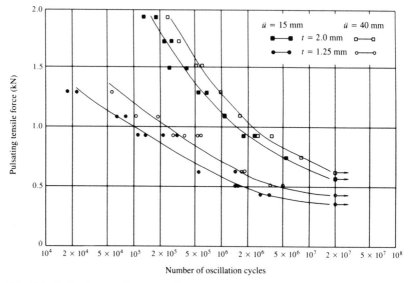

Fig. 5.5   Pulsating tension strength of one-spot specimens as a function of sheet thickness *t* and overlap width *u*

specimens can be expected to exhibit a fatigue strength of 10 per cent of the static shear tension force in the pulsating tensile test, this falls to about 8 per cent with multi-spot specimens. Defects in the weld point, such as pores and blowholes, are virtually insignificant for the failure process under reversal loading, so long as they are in the centre of a weld point with a sufficiently large diameter. The load-bearing behaviour of aluminium spot-weld joints under other types of loading (e.g., cross tension, torsion) is dealt with in detail in (**6**).

5.2.7   Electrode life duration
The electrode life duration is of particular interest in the spot-welding of aluminium alloys. The electrode life is the number of welds performed before quality declines to a minimum of the agreed parameter. This quality parameter can be, in addition to shear tension force, the fracture mechanism (transition from button fracture to shear fracture) or the spot diameter, as well as, for example, the first instance of the electrode adhering to the workpiece, or the surface finish of the welded structure.

As a result of the high contact resistances and the high currents, a lot of heat is produced at contact points between the electrode and the sheet, which favours the mutual alloying tendency of the copper–aluminium materials pair. The consequence of all this is that there is lower electrode life durations with aluminium and aluminium alloys than with steel.

Electrode life performance in spot welding of aluminium alloys is determined by a number of factors, such as the chemical composition, the surface condition, and the thickness of the parts being joined, as well as by the welding parameters (Fig. 5.6). Another important factor affecting electrode life performance is the welding machine used. In addition to mechanical machine characteristics (stiffness, electrode set-down performance, post-setting performance), and electrical machine characteristics (type and form of current), cooling, electrode material and electrode shape all play a part.

Good cooling is important for a high electrode life duration; it facilitates fast heat conduction from the electrode working surfaces and thus should reduce alloying. Direct cooling, in which the coolant passes as

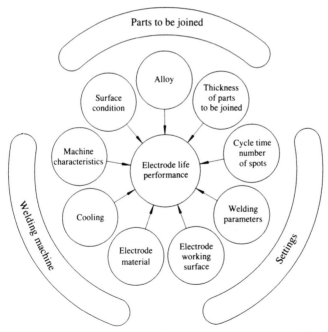

Fig. 5.6 Factors affecting electrode life performance in spot welding of aluminium materials

close as possible to the electrode working surface, is therefore preferable to indirect cooling. Cooling must be more intensive when the spot sequence is larger or the cycle time is shorter.

While both electrodes alloy and wear uniformly during spot welding with alternating current, the anode wears considerably more during welding with direct current. This is due to the so-called Peltier effect, which is illustrated schematically in Fig. 5.7. Due to the relatively large difference in the electrochemical series of about 2 V between the elements Al and Cu, Peltier heat is consumed during the transition of electrons from the Cu cathode to the Al sheet, while Peltier heat is released during the transition from the Al sheet to the Cu anode. The anode thus experiences a greater thermal load, it alloys more severely, and consequently is subject to greater wear. The Peltier effect is greater with a higher flow of current. In the case of steel, this effect is not noticeable due to the smaller difference in the electrochemical voltages and the lower currents. Apart from the uneven electrode wear, the Peltier effect also causes a weld-point displacement in the direction of the anode, which can be utilized when welding sheets of different thicknesses.

In life duration tests with machines with different types of current and different machine designs (Fig. 5.8), greater electrode life durations were achieved with direct current (more uniform energy supply) than with alternating current (high current peaks). The best results were achieved

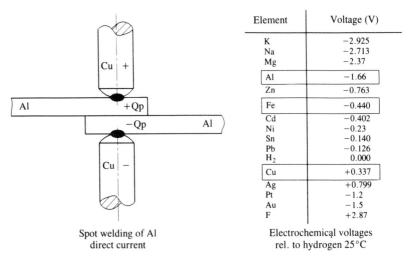

Element	Voltage (V)
K	−2.925
Na	−2.713
Mg	−2.37
Al	−1.66
Zn	−0.763
Fe	−0.440
Cd	−0.402
Ni	−0.23
Sn	−0.140
Pb	−0.126
H_2	0.000
Cu	+0.337
Ag	+0.799
Pt	−1.2
Au	−1.5
F	+2.87

Spot welding of Al
direct current

Electrochemical voltages
rel. to hydrogen 25°C

Fig. 5.7 Peltier effect (schematic)

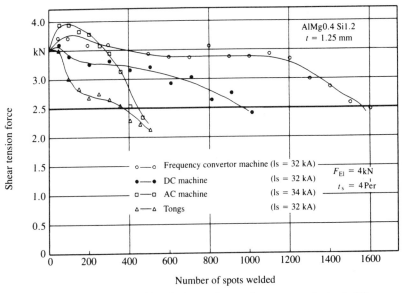

Fig. 5.8   Effect of machine design and type of current on electrode life

with a so-called frequency converter machine, with which the current polarity changes with each weld, and thus both electrodes wear evenly over the whole life duration. When using tongs and mechanically flexible welding machines with a large working radius, the electrode life duration falls considerably compared with that for rigid machines.

5.3   RESISTANCE PROJECTION WELDING

Unlike with spot welding, where the current path is determined by the electrodes, with projection welding the current path is set by the projection (Fig. 5.9). This means that electrodes with large surface areas can be used, which are subject to less mechanical and thermal loading than spot welding electrodes and thus permit greater life durations. In addition, it is possible to weld several weld projections simultaneously using a pair of electrodes. It is true that projection welding of aluminium alloys is made difficult by a number of factors, related primarily to material characteristics.

5.3.1   Weldability

Due to the high thermal and electrical conductivity, the currents required for welding are relatively large. In the case of large surface-area

Resistance

Spot welding   Projection welding

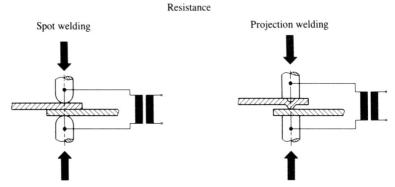

Fig. 5.9   Comparison of spot and projection welding methods. Advantages of projection welding–greater electrode life–several simultaneous welds. Disadvantages of projection welding–large electrode post- adjustments– problems of current and force distribution

parts, this can cause current distribution problems irrespective of the type of machine used.

There are also difficulties resulting from the low mechanical stiffness of the projections and the mechanical/dynamic machine characteristic. When the welding current is actuated there is a danger that the projections will break down very quickly as a result of the high current densities so that, with delayed re-adjustment of the movable electrode, the frictional contact between the electrode and the workpiece is reduced and briefly interrupted. This results in sweat contact and spatter which impairs the quality of the welded bond and damages the electrodes (7).

As with spot welding, with projection welding the high contact resistance caused by the oxide film is an important factor. The pre-treatment methods described for spot welding also apply to projection welding. When using untreated sheet, the expected scatter in strength values is lower with projection welding than with spot welding due to the higher area pressure from the projection.

5.3.2   Parameter settings

The choice of welding parameters is dependent on the geometry of the projection, as well as on the factors familiar from spot welding. Current levels for spot welding can be used as guidelines. Basically both alternating (a.c.) and direct current (d.c.) can be used for welding. When using a.c., the range of welding parameters which can be set is smaller. With projection welding, the electrode facing the sheet with the projection is subject to considerably greater wear due to melting of the projection. On the other hand, the anode is subjected to greater loading with d.c. welding due to the Peltier effect. To counteract uneven electrode wear in the case of machines with constant polarity, the part with projections should be in contact with the cathode where the parts to be joined are the same (material, thickness) (8).

During the rate time the electrode force must adapt itself to the mechanical stiffness of the projection geometry in question. Electrode force and projection geometry should be matched so that about 60 per cent of the projection height is maintained when the electrode force is applied (9). As with spot welding, relatively short current times are necessary for the projection welding of aluminium alloys.

Large surface-area, flat electrodes made of CuCrZr are usually used for projection welding. With these electrodes, the faces must be precisely parallel. In many cases it may be advisable to work with ball electrodes (large ball radius).

5.3.3   Projection geometry

Figure 5.10 summarizes the conventional types of projection for projection welding. With embossed projections, a difference is drawn between free stamping and mould stamping. In free stamping the projection is produced using a stamp with perforated plate, and in mould stamping it

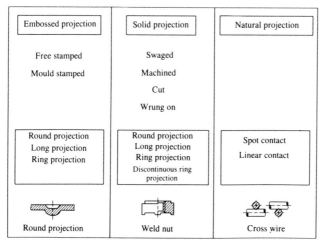

Fig. 5.10 Types of projection for projection welding

is produced with a stamp and a die. With aluminium materials, mould stamping has the advantage of strengthening the material, which has a favourable effect on the stability of the projection before the welding current is applied. Round, long and ring projections can be produced with both processes. The long projection is not suitable for aluminium due to its low level of rigidity. Ring projections are suitable for sheet up to 2.5 mm thick, and round projections can be used with sheet 2 mm or more thick. The use of solid projections is appropriate mainly for parts whose manufacture facilitates the simultaneous formation of one or more projections, such as, for example, with weld nuts. Solid projections can also be created in sheet materials with material forming similar to reverse extrusion using a profiled stamp.

Although extensive research work has demonstrated that projection welding can be used successfully to join aluminium parts, few industrial applications are known to date.

5.4 INERT GAS METAL ARC WELDING

5.4.1 TIG welding

Of the inert gas welding methods, manual TIG (Tungsten Inert Gas) welding is the most important for joining aluminium body materials, as it is particularly suitable for thin sheet.

In TIG welding, the electric arc burns under an inert protective gas between a tungsten electrode which does not flash off and the workpiece; the workpiece and possibly the filler metal is fused. Most metals are welded with direct current. Here the tungsten electrode is negatively polarized, to prevent thermal overload and thus the destruction of the extremely hot electrode. The protective gas which protects the tungsten electrode and the melt bath against oxidation is usually argon.

Due to the high melting point of the oxide film, aluminium materials cannot be welded under argon with a negatively polarized electrode. Only when a positively polarized electrode is used is the oxide skin ruptured and destroyed. This desired 'cleaning effect' with positive polarity of the electrode occurs due to the fact that only the positive gas ions have sufficient kinetic energy to perforate the oxide skin of the surface of the workpiece (Fig. 5.11). When helium is used, aluminium can be TIG welded with direct current and a negatively polarized electrode. Here, however, the arc length must be kept very constant, so this method is not suitable for manual welding.

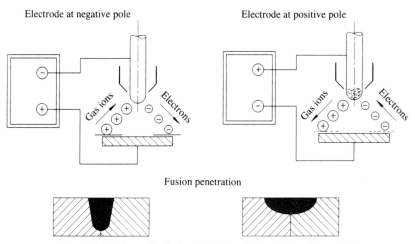

Fig. 5.11 **Effect of polarity in TIG welding of aluminium** (10)

As positive polarity would subject the electrode to thermal overload, aluminium and its alloys are generally welded with alternating current. If, in exceptional cases, positive polarity has to be used, it should be noted that the current-bearing capacity of the electrode is lower than with a.c. welding by about a factor of four.

Welding with a.c. is a good compromise with regard to a good cleaning effect, high current-bearing capacity and long electrode life. During the change of polarity, the oxide skin ruptures during the positive half-wave and cools the electrode during the negative half-wave.

Depending on current loading, the diameter of the tungsten electrode is 1.6 to 6 mm. Whereas tungsten electrodes containing 1 to 2 per cent thorium oxide are used for d.c. welding, pure tungsten electrodes are used almost exclusively for a.c. welding of aluminium. The electrodes are not sharpened for welding aluminium. If the electrode has the appropriate diameter for the current, the ideal shape of a hemisphere forms automatically at the end.

TIG welding is suitable for all welding situations. In the sheet thickness range up to 3 mm, welds are made with no gap between the parts being joined, for all types of seam. Guideline values for manual downhand TIG welding (PA (ISO 6947)) are given in Table 5.5. For vertical and overhead welding positions (respectively PF and PE (ISO 6947)) the welding current is about 20 to 40 A lower; the welding speed is reduced

Table 5.5   Approximate values for manual TIG welding of aluminium— using alternating current, downhand welding

Type of seam	Thickness of sheet (mm)	Welding current (A)	Electrode diameter (mm)	Argon consumption (l/min)	Welding velocity (mm/min)
Edge-formed	1.0	40–70	1.6	4.5	380
1 weld	1.0	50–60	1.6	4.5	400
	1.5	60–80	1.6	5.0	350
	2.0	80–100	2.4	5.5	320
	2.5	100–130	2.4	6.0	300
	3.0	120–150	2.4	7.0	280
Fillet weld	1.0	60–70	1.6	4.5	400
	1.5	70–90	1.6	5.0	350
	2.0	90–120	2.4	5.5	320
	2.5	110–150	2.4	6.0	300
	3.0	140–170	2.4	7.0	280

accordingly. With manual welding, the filler metal used consists of bars 1 m long. The choice of filler metal is dependent on the material of the parts being joined or on the combination of materials being joined. Body materials are generally welded with S-AlMg5.

The electrode must not touch the workpiece during ignition or welding, as it could cause defects in the weld due to the inclusion of tungsten particles. This requirement has led to the development of igniters which permit non-contact ignition. In the case of a.c. TIG welding, there is also the fact that the arc is extinguished at each zero crossing. Automatic re-ignition of the arc in the low-energy initial phase of each half-wave is not possible. Due to the temperature drop resulting from heat conduction and the absence of easily-ionizable elements, the residual ionization in the arc path is too low. For this reason, high-voltage pulses must be superimposed on the voltage after each zero crossing, which permit half-wave ignition and thus a stable electric arc.

Re-ignition is affected by the following, in addition to the igniter: welding parameters, material temperature, melt (oxide), tungsten electrode (composition, shape), welding torch, and filler metal.

The combination of aluminium (workpiece) and tungsten (electrode) via the arc as an electrical conductor produces a 'rectified' welding current. Electron emission and thus current flow depend on the temperature pertaining at the cathode. Therefore, in the positive half-wave (electrode with positive polarity), less current flows than in the phase with a negative-polarity tungsten electrode (melting temperature approximately 3300°C) owing to the low melting temperature of the aluminium. The negative d.c. component produces a small cleaning effect, a hard and unstable arc, and a deep fusion penetration. Power sources with 'balancers' make it possible to counteract the differing electron emissions of aluminium workpiece and tungsten electrode (**11**). The current/time areas of the positive and negative a.c. components can be preselected and controlled during welding. This control system can influence:

- the cleaning effect, relative to the surface state of the workpiece;
- the fusion penetration depth and weld seam geometry;
- heat penetration into the workpiece;
- the metallurgical quality of the weld and its boundary zones;
- compatibility with different alloys.

5.4.2   MIG Welding

In MIG welding, the arc burns between the workpiece and the fusing wire electrode, which is at the same time the welding filler. Welding is performed with d.c. with a positive-polarity electrode. The high thermal loading of the wire electrode is desirable and increases the fusion capacity. However, this welding method has certain serious disadvantages compared with TIG welding for joining aluminium body sheet, which impede increased utilization for making bodyshells.

As the arc must burn continuously to produce the necessary cleaning effect, the short arc-welding method, which facilitates the welding of thin body sheet due to the lower currents, cannot be used in MIG welding of aluminium, unlike in MIG welding of steel. To achieve short-circuit-free drop transition in the spray arc, a critical current value, which is dependent on the wire electrode diameter selected, must be exceeded. Due to this minimum current value, MIG welding of aluminium with wire electrodes of conventional diameter (0.8–2.4 mm) is not suitable for thin sheet.

If MIG welding is to be used in spite of this, it is possible to use thinner wire electrodes or MIG pulsed-arc welding. Thinner wires can cause wire feed problems, and the pulsed-arc method requires special equipment. In pulsed-arc welding, the energy source is periodically switched between two characteristic curves. This means that a low basic current and a higher pulsed current flow alternately. The pulsed current must be above the critical current level, and so the range of the spray arc and thus short-circuit-free material transition is reached. As the high current flows only in the pulse phase, this process is suitable for thin sheet.

5.5 LASER-BEAM WELDING

Laser-beam welding is becoming increasingly important in body construction owing to its process-specific advantages. The most important advantages are:

- lower heat production;
- less component distortion;
- non-contact welding;
- high welding speed;
- high capacity for automation;
- high flexibility;
- higher component rigidity than with spot-welded components.

In addition to the disadvantages familiar from laser-beam welding of steel sheet, such as high investment costs and high demands on accuracy, the specific thermal and visual material characteristics hinder laser-beam welding of aluminium (12). To achieve adequate weld depth, material-related threshold intensity of the laser radiation is necessary. Owing to the low absorption and high thermal conductivity of aluminium, a higher threshold intensity is necessary than with steel. This assumes a high-output laser and focussing to as narrow a beam as possible. On the other hand, this high radiation intensity combined with the lower ionization energy compared with steel leads to increased plasma formation, which can cause the workpiece to be screened from the laser radiation. The parameter range within which weld seams can be produced with reproducible high quality is therefore more restricted than for laser beam welding of steel. The welding process is more sensitive and requires all parameters to be well matched (13). The working gas is particularly important here with regard to composition, flow rate and gas jet geometry and arrangement. Unlike all other welding methods, the oxide film on the workpiece surfaces is useful, as it lowers the necessary intensity threshold owing to its better absorption.

In addition to the beam and process parameters, chemical composition and the state of the parts to be joined have an important effect on the weld quality, especially with regard to porosity, undercuts and hot crack formation.

Despite promising research work, laser-beam welding of aluminium, unlike laser-beam cutting, has not gained access to industrial manufacturing.

5.6 PENETRATION BONDING

Methods which are combined under the heading of penetration bonding are among those joining methods for body construction which are gaining in importance. Three variants of this method of mechanical joining are illustrated in Fig. 5.12. We differentiate between single-stage ('Pressure Joining'®) and multi-stage (clinching) penetration bonding, in both of which a cutting process takes place, and one-stage 'TOX Joining'®, in which extrusion is performed. The typical joint geometry for 'TOX Joining'® is the round shape, while for the other methods star shapes and cross shapes are possible in addition to beam shapes. A fourth type, not shown in the illustration, Tog-L-Loc, is comparable with 'TOX Joining'®. Unlike with TOX Joining®, the mould which is usually in three parts opens during the joining process. This improves the flow behaviour of the material.

The advantages of penetration-bonding methods compared with resistance spot and projection welding lie in the comparatively low plant costs and inexpensive installation and power consumption. Despite the higher tool costs for joining aluminium sheet, the manufacturing costs for the bonding processes are generally considerably lower than for the two resistance pressure welding processes, owing to the greater tool life durations. As they are mechanical joining processes with no heat production, there is no welding spatter or thermal distortion. It is not necessary to pretreat the surfaces to remove the oxide film as with the resistance welding processes. Furthermore, it is possible to use these joining methods to bond aluminium and steel together, which is not possible with welding. The possibility of contact corrosion must be taken into account here, however.

The lower strength under static loading is a disadvantage compared with spot-welded joints. In addition, the appearance of the joint can be unattractive, so it is frequently impossible to use it in a visible location. While, in the case of two-stage clinching, as with spot-welding, it is possible to join sheets of different thicknesses and material combinations with the same set of tools by varying the bonding parameters, with all

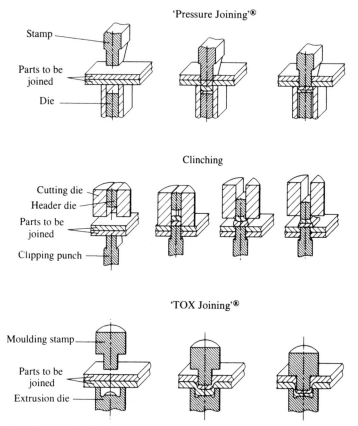

Fig. 5.12   Standard tools and process sequences in penetration bonding (14)

other penetration-bonding methods it is necessary to have different combinations of stamps and dies.

To study the load-bearing capacity, the shear tension force determined in a shear tension test is also usually used for penetration bonding. Depending on the type of loading, the cross tension test and the peel test may also be used. The highest shear tension forces are possible with the cross clinch with aluminium materials (**15**). However, only 80 per cent of the shear tension strength of a 5.5 mm diameter weld spot is reached. The shear tension force of the beam clinch is dependent on position. The best results are achieved if the beam clinch is transverse to the direction of the load. Compared with the cross clinch, shear tension strength falls to about 40 or 50 per cent. With the 'Pressure Joining'®bonding spot, which is also location-dependent, the shear tension forces are somewhat greater than with the beam clinch. The 'TOX'® round spot has, depending on the spot diameter used, about 50 to 80 per cent of the shear tension force of a weld spot with a diameter of 5.5 mm (16). The cross tension and peel forces which can be transmitted are relatively small. Under dynamic loading, all penetration-bonding spots exhibit good strength performance. The fatigue strength of comparable spot-welded joints is reached and sometimes even exceeded.

In the case of penetration-joining methods which involve predominantly cutting ('Pressure Joining'®, clinching), the surface state (as-delivered state, oiled, chromated, painted) does not have an important effect on the strength of the bond (17). Conversely, with the other penetration-bonding methods, oil or paint films on the surface of the parts to be joined have a negtive effect on the bond quality, while chromating considerably increases the strength. Furthermore, the quality of penetration-bonded joints is also affected by variations in mechanical sheet characteristics, by sheet-thickness tolerances, by variations in parts fit, and by tool tolerances and wear (18).

5.7 MANUFACTURE OF AN ALUMINIUM ROOF FOR A ROADSTER

In order to mechanize the production of a roadster roof made of aluminium, a new manufacturing concept was developed in the form of a flexible manufacturing cell. The roof consists of ten separate parts. AlMg0.4Si1.2 is used for the planking and AlMg5Mn for the interior parts, both types of sheet being 1.25 mm thick. The roof requires 448 resistance spots and about 400 mm of TIG weld seam. The overall layout of the plant, Fig. 5.13, is divided into four work planes:

- material preparation;
- insertion;
- welding;
- component handling.

Insertion of the separate parts is performed manually at stations 1 to 4 (insertion level). On the welding level, the spot weld joints are produced with three overhead robots at stations 1 to 4 and with a floor-mounted

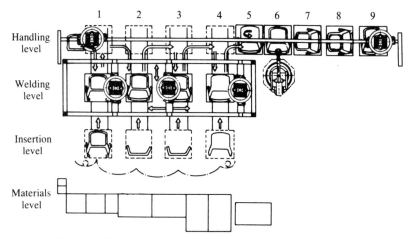

Fig. 5.13   Flexible manufacturing cell for an aluminium roof for a roadster

robot at station 6. At station 5 the component is turned and then seam-folded at stations 7 and 8. The individual roof assemblies are transported within stations 1 to 4 by rail-mounted lift trucks. Between stations 1 and 9 the components on the handling level are moved by two robots on a linear gantry.

This plant design provides the following different types of flexibility:

- flexibility for modification;
- model flexibility;
- successional flexibility;
- flexibility in production quantities (up to the maximum production capacity).

The components are welded in the etched state using a medium-frequency welding frequency-converter installation, consisting of a frequency converter, a transformer and a rectifier. The welding frequency converter technology facilitates the use of robot welding tongs with integral transformer and rectifier, despite the high welding currents of up to 34 kA, as the higher frequency permits the size and weight of the welding transformer to be reduced. Direct current was chosen in view of the smaller inductive losses, the symmetrical supply system loading, the better weld quality, and the higher electrode life duration compared with alternating current. The electrode caps are replaced automatically after about 400 welding operations. The TIG welds are performed manually.

6

Solutions to problems with extruded sections in vehicle construction

J. Maier and F. Wehner

6.1 OBJECTIVE

Environmental considerations and rising energy costs promote the development of vehicles which use less petrol. Reducing the weight of the body and the chassis makes a significant contribution to this aim. The use of aluminium extruded sections can also bring about cost savings in vehicle production.

6.2 THE EXTRUSION PROCESS

Extrusion is a typical hot-forming process, in which the desired cross-sectional shape is produced in one single forming stage.

Figure 6.1 shows the principle of extrusion. The material to be used is a continuous casting billet heated to about 500°C. A hydraulic medium (oil or water) is used to drive the ram towards the die. The billet enclosed in the receiver is upset, and the material flows through the die opening as section, rod, or tube. The continuous billet, up to approximately 40 m long, produced in this way, is cooled with water or a blower to room temperature, drawn out, and then cut to the desired length.

As every catalogue of sections and every section design shows, there are virtually no limits to section profiling in aluminium. In DIN 1748, Part 3, a differentiation is made between solid sections, semi-hollow sections and hollow sections. Solid sections have a simple geometrical shape without enclosed surfaces; semi-hollow sections include all sections of more complex shape which do not have completely enclosed surfaces. Hollow sections have one or more completely enclosed chambers.

Solid sections and semi-hollow sections are extruded with smooth or flat dies. These are plates made of hot-forming steel, in which the shape of the desired section is created with wire erosion and mechanical precision finishing (Fig. 6.2). The whole tool set also includes auxiliary tools;

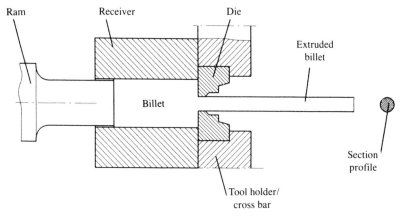

Fig. 6.1   Direct extrusion, schematic

two-part tools are necessary for hollow sections (Fig. 6.3). The die plate shapes the outer contour as with a solid section; a mandrel, supported on the die plate via brackets, shapes the internal contour. The hollow section is extruded in two stages: first the bar is rammed in several billet sections into the inlet of the mandrel part. The sections of billet can join up again in the welding chambers under the brackets. The section is then shaped in the second stage.

Hollow sections have continuous extrusion seams at locations where

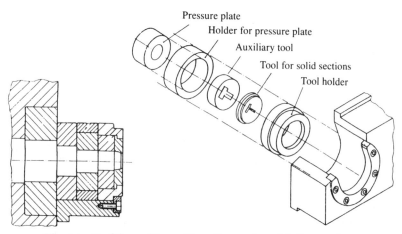

Fig. 6.2   Tool for making solid sections and semi-hollow sections

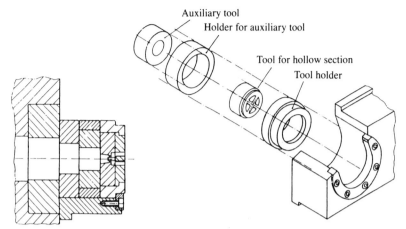

Fig. 6.3   **Tool for making hollow sections**

the flow components in the tool meet to give pressure welding. Extrusion seams should in no way be equated with a pressure weld; if the extrusion parameters are correctly chosen, they are only visible after etching as crystallographic differences in microstructure. Furthermore, the microstructure form can cause a decline in the extension values with tension members transverse to the extrusion seam.

6.3   EXTRUDED SECTIONS AND MATERIALS

The extrusion sector is dominated quite definitely by hardenable materials, which gain their strength from heat treatment: solution heat treatment (usually combined with the extrusion process), fast cooling from solution heat treatment temperature, and artificial ageing. The hardenable alloys span a range from F22 to about F70; in the automotive industry, medium-strength AlMgSi alloys are primarily used, the most important examples of which are shown in Table 6.1. It is possible to disregard high-strength materials, as dimensioning in automotive development is directed mainly towards rigidity criteria and crush behaviour. Rigidity is determined by the modulus of elasticity. The aluminium materials generally used exhibit the same modulus of elasticity of 70 000 N/mm^2.

Crush behaviour determines additional requirements of the designer, such as, for example weld seam strength (Table 6.2). Aluminium materials react to the welding heat; along a heat-affected zone about

Table 6.1   Aluminium alloys tested in vehicle construction and their strength values

Designation + state		Minimum strength values		
DIN	ISO	$R_{p0.2}$ (N/mm^2)	R_m (N/mm^2)	A_5 (%)
Al99.9MgSi	–T6	195	240	14
AlMgSi0.5 F22	6060–T6	160	215	12
AlMgSi0.5 F25	6060–T6	195	245	10
AlMgSi0.5Mn F25*	6106–T6	200	250	10
AlMgSiCu F26*	6061–T6	240	260	9
AlMgSi0.7 F27	6005A–T6	225	270	8
AlMgSi1 F31	6082–T6	260	310	10
AlZn4.5Mg1 F35	7020–T6	290	350	10

* not currently standardized in DIN 1748

20 mm wide on both sides of the weld, $R_{p0.2}$ falls to about 50 per cent of the base material (Fig. 6.4).

AlZnMg materials, which harden again after welding, are an exception. The loss of strength during welding must be compensated for, if necessary, by appropriate shaping (such as local thickening, see Fig. 6.5) or by moving the weld seam to areas of the component which are subject to lower stress.

Table 6.2   Strength of welded joints (transverse tension members)

Alloy	MIG butt weld*			TIG butt weld†	
	$R_{p0.2}$	R_m	(N/mm^2)	$R_{p0.2}$	R_m
AlMgSi0.5 F22	80	110		65	95
AlMgSi0.5Mn F25‡	105	165		–	–
AlMgSi0.7 F27	115	165		100	150
AlMgSi1 F31	125	190		110	170
AlZn4.5Mg1 F35§	205	275		205	275

* wall thickness <10 mm
† wall thickness <6 mm
‡ not currently standardized in DIN 1748
§ strength after 3 months at R_t, recommended for use only in the artificially aged state, filler S-AlMg4.5Mn

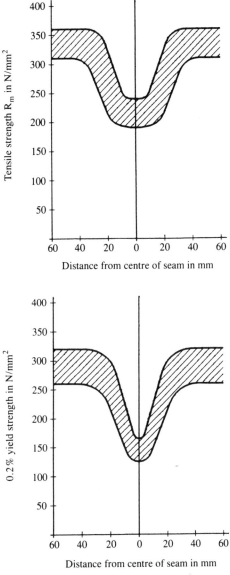

Fig. 6.4 **Strength values for the heat-affected zone in the area of the weld seam for AlMgSi1F31 (longitudinal tension members, wall thickness ≤15 mm)**

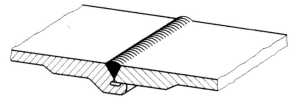

Fig. 6.5   Wall thickening of aluminium sections to compensate for strength loss in the heat-affected zone, with additional weld pool protection

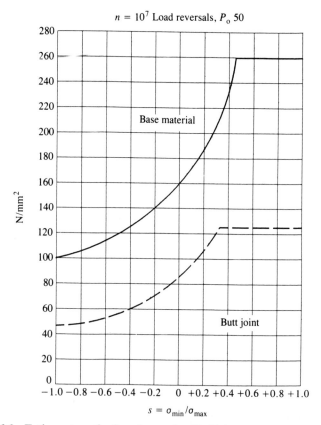

Fig. 6.6   Fatigue strength of sections and welded joints made of AlMgSi1F31

Where cyclic loading is concerned, fatigue strength data are of interest. Moore-Kommers-Jasper diagrams exist for all alloys; Fig. 6.6 gives as an example the permissible stresses for 10^7 load reversals in the various stress states. The curve for the base material and the horizontal branches at high S-values (protection against $R_{p0.2}$ and R_m) are influenced by the material. The curve for welded joints is largely identical for all materials. Here the notch factor and micro-defects in welding predominate.

There are other characteristics which can be significant for the selection of the material for a section.

(1)   A decorative appearance after anodizing is achieved by choosing low-alloy AlMgSi materials of high base purity, such as, Al99.9MgSi or AlMgSi0.5.

(2)   Good corrosion resistance is assured with all AlMgSi materials. Higher-strength materials of the AlZnMg family require a protective coating, especially in the heat-affected zone of welded joints.

(3)   Other characteristics such as bending capacity, conductivity, thermal resistance, wear, fracture toughness, etc., depend to such an extent on the individual case that further discussion on this remains the subject of technical agreement between the user and the manufacturer.

6.4   FORMS IN WHICH SECTIONS CAN BE SUPPLIED, AND ECONOMIC CONSIDERATIONS

Extrusion technology is widely available in Europe. In western Europe (14 countries) there are currently 458 extrusion plants with a total capacity of about 1.6 milion tonnes. The size of the profile cross-section which can be manufactured is measured by the circumscribed circle and depends on the available press force and the diameter of the receiver. The west European presses have

- press forces of about 500–10 000 t;
- receivers of about 80 mm to 670 mm ϕ or 750 × 260 circumscribed rectangle;
- capacity for sections with a circumscribed circle of about 10 mm to 600 mm ϕ or 800 × 100 mm circumscribed rectangle;
- weights per metre ranging from a few g/m to > 100 kg/m.

No over-engineering

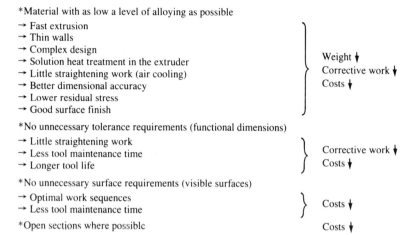

*Material with as low a level of alloying as possible
→ Fast extrusion
→ Thin walls
→ Complex design
→ Solution heat treatment in the extruder
→ Little straightening work (air cooling)
→ Better dimensional accuracy
→ Lower residual stress
→ Good surface finish

Weight ↓
Corrective work ↓
Costs ↓

*No unnecessary tolerance requirements (functional dimensions)
→ Little straightening work
→ Less tool maintenance time
→ Longer tool life

Corrective work ↓
Costs ↓

*No unnecessary surface requirements (visible surfaces)
→ Optimal work sequences
→ Less tool maintenance time

Costs ↓

*Open sections where possible

Costs ↓

Fig. 6.7   Economic considerations for the section manufacturer

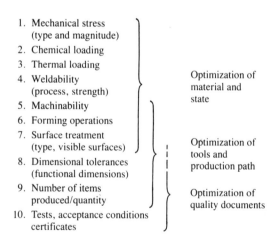

1. Mechanical stress
   (type and magnitude)
2. Chemical loading
3. Thermal loading
4. Weldability
   (process, strength)

Optimization of material and state

5. Machinability
6. Forming operations
7. Surface treatment
   (type, visible surfaces)

Optimization of tools and production path

8. Dimensional tolerances
   (functional dimensions)
9. Number of items
   produced/quantity

Optimization of quality documents

10. Tests, acceptance conditions
    certificates

→ Conduct discussions with the section manufacturer at the development stage

Fig. 6.8   Check list for the designer

The minimum wall thickness depends on material, section type, section shape and size. For small- to medium-sized AlMgSi0.5 sections, the minimum wall thicknesses are about 1.2 to 1.8 mm. In automotive construction, wall thicknesses are generally not < 1.0 mm for reasons of stiffness (exception: heat exchangers and decorative trim).

Section tolerances and guidelines for section design are given in DIN 1748, Parts 3 and 4. In addition, the earliest possible contact with the manufacturer of the sections is advisable, so that all production advantages and cost-saving potentials are fully utilized.

A vehicle builder who is familiar with the tool costs of sheet-drawing tools finds the manufacturing costs of extrusion dies unusually low: they range from about DM1000 for simple small sections up to about DM80 000 for extremely large sections. A range of DM3000 to DM10 000 can be assumed for applications in the automotive sector (fixed costs).

Figure 6.7 lists some economic considerations for the section manufacturer: easier extrusion at the section manufacturing plant leads finally to savings of weight, refinishing and corrective work, which reduces running costs. The check list in Fig. 6.8 is an aid to minimizing costs in agreement with the section manufacturer.

6.5 PROCESSING OF SECTIONS

The extruded sections are long and straight. To produce components ready for use, subsequent processing stages are necessary, consisting of cutting, joining, forming and/or surface treatment.

The advantages of light alloy come to the fore in machining: aluminium materials can be dealt with much more easily and more rationally than steel. Short, cut pieces of section are frequently an alternative to forgings.

Bonding and surface treatment are dealt with in other chapters. Table 6.3 gives an assessment of the joining methods available today for joining sections to one another and to other semi-finished products. In the future, adhesive bonding development will have to continue and will require more intensive investigation. The same applies to laser welding, which, at the present stage, can be used for wall thicknesses up to about 6 mm.

What drawing is to sheet, bending is to sections. When bending aluminium materials, greater resilience must be expected than with steel,

Table 6.3   Assessment of familiar methods for joining aluminium sections to one another and to other semi-finished aluminium products

	Resistance spot welding	TIG MIG	'TOX Joining'® 'Pressure Joining'®	Bolting Riveting	Adhesive bonding
Static strength	+	+ +	0	+ +	+ +
Dynamic strength	0	+ +	+	+	+ +
Creep behaviour	+	+	+	+	−
Crash behaviour	+	+	+	+	+ +
Corrosion performance	+	+	+	0	+ −
Technical development level	+ +	+ +	+	+	0
Cost	0	↑	↓	↑	↑ ↑
Example	← Audi 100 →		Rover	Mercedes Benz	Proto- types
	Door window frame Section/ sheet	Section/ section	Roof Section/ sheet	Seat rail Section/ Fe/Sheet	

+ + = very good; + = good; 0 = neutral; − = limited applicability; ↑ = higher costs; ↓ = lower costs

Table 6.4   Bending factors for extruded sections

DIN designation	Bending factors, f_w, for state		
	0 Soft	T4 Cold age hardened	T6 Artificially aged
Al99.9MgSi		4.0	5.5
AlMgSi0.5(Mn) AlMgSi0.7		4.8	6.2
AlMgSi1 AlMg1SiCu	1.5	5.0	7.5
AlZn4.5Mg1	2.5	5.5*	9.0*

* only recommended with subsequent heat treatment

owing to the lower modulus of elasticity, i.e., the section must be bent to a greater extent. The elongation of the outer fibres is important for the deformation limit; this depends on the hardening state of the section material and also on the bending angle and the internal radius. There is an approximation formula for the smallest possible bending radius r_i:

$$r_i = f_w(0.8d - 2)$$

f_w being the bending factor and d the section height $(d > \approx 12$ mm). Table 6.4 gives values for f_w. As strength declines, both for the material and the state, deformability increases. In certain cases it can be advisable to perform the bending work in the cold precipitation-hardened state or even in the soft state, and to produce the full strength in the material by means of subsequent heat treatment. To ensure a uniform shape with the generally larger radii and with section profiles which are not simple, stretch-bending methods must be used, especially for biaxial bending. Stretch-bending tools are expensive compared with extrusion dies, but much cheaper than drawing tools for sheet.

6.6 AUTOMOTIVE APPLICATIONS OF SECTIONS

Often it is the specific characteristics which make the choice of an aluminium material appropriate. Here, the low weight is certainly a prime characteristic. In recent years, however, a factor other than physical and chemical properties has brought the decision down on the side of aluminium, namely the outstanding forming capacity of extrusion moulding. The high level of design freedom permits 'intelligent sections' with cost advantages.

In the case of 'intelligent sections', functions are integrated so that material, weight, corrective processes and finally expense are reduced. The fundamental opportunities for building functions into sections are:

- surface design (e.g. appearance, cooling);
- spot joins for assembly and fastening (e.g. bolt channels, 'T' grooves);
- linear joints (e.g. groove and spring, clips, hinges, insetgrooves);
- local reinforcement (e.g. welding edges, webs, bolt points);
- static load-bearing elements (e.g. internal webs, reinforcing flanges);
- miscellaneous functions (e.g. seals, stops, ducts, cable channels, preferred buckling points).

Figures 6.9 and 6.10 give examples of 'intelligent sections', which can reduce costs and time-consuming operations.

The list of past and current applications in vehicle construction is diverse (and incomplete). All parts in Table 6.5 are/were in production; numerous others are in development. The benefit afforded by 'intelligent sections' is illustrated below with some examples.

(1) The knee bolster tube (knee bar) prevents submarining by the front passengers in a crash (Fig. 6.11). A copy of the previous steel version, namely two tubes welded with various flanges and reinforcing plates, would indeed be lighter, but also considerably more expensive. Instead of the differential steel part, an integral aluminium part was developed, in which the section provides the functions of buckling stiffness by means of interior webs, strengthening points by means of bolt channels, and mounting for the padding by means of lateral webs. The integral elements of the shape are not work- and cost-saving ballast material, but are statically load-bearing and thus support the main function. Figure 6.12 shows that the aluminium design has a distinct advantage in crush performance.

(2) A seat rail made of steel sheet consists of canted sheet parts. That the aluminium seat rail (Fig. 6.13) is lighter, better and neutral with regard to cost is due to the fact that the section not only performs

Table 6.5   Examples of applications for sections in vehicle construction

Window frames	Audi 80/100
Window lifter rails	
Seat rails	Mercedes, BMW, Porsche
Bumpers	Mercedes, BMW, Porsche
Roof drip mouldings	Golf, DB W201, DB W124
Door crush member	Audi 80/100 USA
Trim	All
Roof rail	VW Passat, VW Polo, Mercedes, Audi, Opel, etc.
Brake distributor housing	
Steering articulated shaft	Elbe
ABS–housing	Bosch
Fuel collector pipe	
Radius rods	Mercedes
Propeller shaft	BMW
Knee bar	

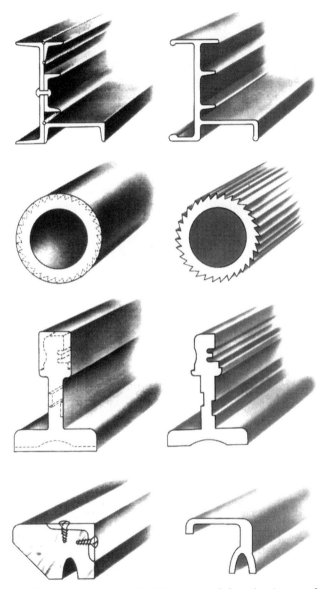

Fig. 6.9   Cost savings by using 'intelligent extruded sections', examples (1)

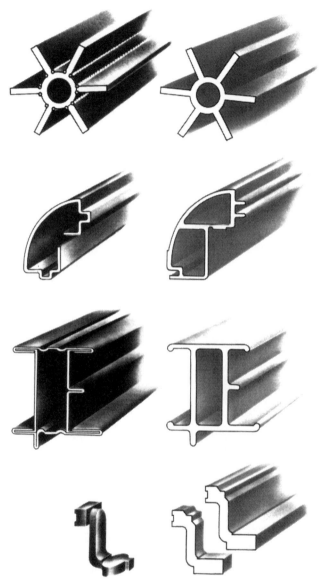

Fig. 6.10   Cost savings by using 'intelligent extruded sections', examples (2)

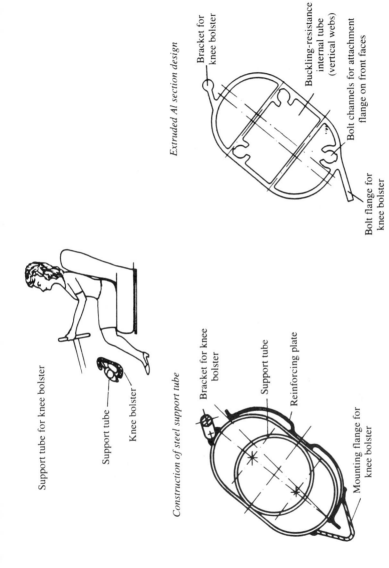

Support tube for knee bolster

Support tube

Knee bolster

Construction of steel support tube

Bracket for knee bolster

Support tube

Reinforcing plate

Mounting flange for knee bolster

Extruded Al section design

Bracket for knee bolster

Buckling-resistance internal tube (vertical webs)

Bolt channels for attachment flange on front faces

Bolt flange for knee bolster

Fig. 6.11   Knee bolster tube in steel and aluminium

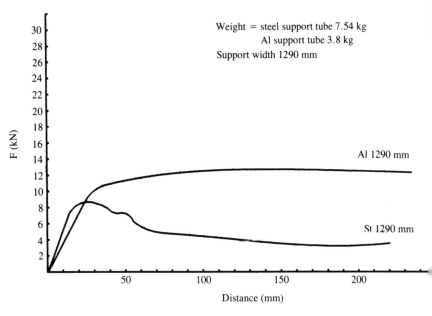

Fig. 6.12   Comparison of crash performance of welded steel design and extruded aluminium section version of knee bolster tube

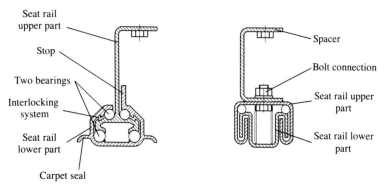

Fig. 6.13   Passenger car seat rail–comparison of steel and aluminium designs

roller guidance and stopping, but also absorbs considerably higher fracture loads in a crash by means of an interlocking structure, without the necessity for thicker walls.

(3) The side impact member in the doors protects passengers in an impact in the door region. As an alternative to steel tubes, an aluminium-section version was developed (Fig. 6.14) which gives an optimal force–displacement response owing to the cross-sectional design and deliberately positioned notches. In addition to the main function of safety, its advantages are: lower weight, simple assembly, and good corrosion resistance in the poorly ventilated interior of the door.

Extruded aluminium sections are used not only as individual components or a system of sections; it is often useful to combine them with other types of semi-finished products or with castings.

– Section with sheet: drawn door assembly member with sectional window-frame; roof skin with extruded rain channel.

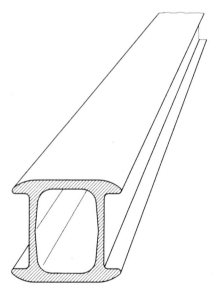

Fig. 6.14 Aluminium section for side impact member

- Section with forging: roof rail with extruded gallery bar and forged support feet (sometimes also cast); steering articulated shaft consisting of sectional pieces, forged parts and cold extruded parts; articulated shaft made of tubes with friction-welded forged flanges.
- Section with casting: support tube with cast flanges; seat frame of tubes, sheet brackets and cast junctions.

In future, the use of sections will increase for economic reasons if the space-frame design proves successful. This is a sectional frame structure, with weldable cast junctions. Figure 6.15 shows the structure of the aluminium-intensive vehicle (AIV), a joint development by Audi and Alcoa. With 70 kg of sections and 70 kg of castings, the design is about 40 per cent lighter than a steel version. The lower tooling costs mean considerable savings for serial production (Fig. 6.16).

To summarize, the following can be stated.

- Extrusion has reached a high technical standard and allows the designer a varied range of section profiles.

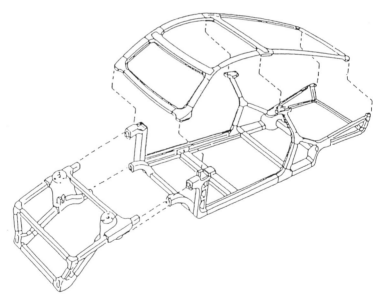

Fig. 6.15   Space-frame design made of extruded sections and cast junctions (from McClure)

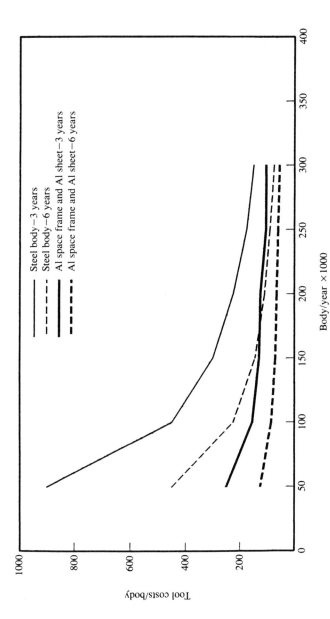

Fig. 6.16 Tooling costs for steel bodies and aluminium space frames as a function of number produced and amortization time (from McClure)

- Materials development creates strong, corrosion-resistant materials which have proved themselves in practice.
- By integrating functions, 'intelligent sections' can be designed, which save materials, weight, corrective work and costs.
- In future, extruded aluminium sections will find additional applications in vehicle construction, because the benefit deriving from weight reduction, corrosion reduction, simple processing and ease of recycling is convincing.

Aluminium – new casting methods and materials properties

K.-H. v. Zengen

7.1 INTRODUCTION

The catchword 'near net shape parts' has been part of our vocabulary for only a few years. However, the idea behind it – namely the use of structural materials in a way which is economical with resources and requires little processing – has always been applied in the manufacture of castings. From the liquid metal to a blank in only one production stage–that guarantees a high level of economy. Furthermore, no other metal has so many different possible casting methods as aluminium. So it is not surprising that of approximately 55 kg of aluminium used today in the average passenger car, more than 80 per cent is used in the form of castings. Only the lack of ductility, which is a disadvantage of cast materials in general, appears to stand in the way of more widespread use.

Nevertheless, developments made in recent years have brought marked improvements and will surely help to open up new applications for aluminium.

7.2 SYSTEMATICS OF CASTING METHODS

The choice of the 'right' casting method is influenced by a number of factors (Fig. 7.1). Casting methods are divided basically into those using 'lost' moulds, and those with 'permanent' moulds (Fig. 7.2). The best-known representative of the first type is sand-casting, and the second heading covers mainly pressure die-casting and chill-casting.

Almost all classical casting methods produce components which – at least sometimes – have porosities, which arise either from gas inclusions (including air) or from volume contraction during solidification with inadequate after-pouring. It is precisely porosity, however, which has a negative effect on mechanical properties, in particular fatigue strength, in too high a percentage of cases (1). Gas inclusions make heat treatment

★ Size of casting
★ Shape of casting/level of difficulty
★ Total production quantity
★ Daily/weekly production quantity
★ Development period
★ Development risk
★ Dimensional accuracy required
★ Toughness required

Fig. 7.1   Criteria for selection of a casting method

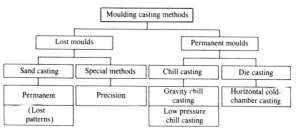

Fig. 7.2   Summary of general mould-casting methods for aluminium

(solution heat treatment) impossible, which means that the strength potential of many cast materials cannot be fully utilized.

The development of new casting methods is therefore directly linked to the avoidance/minimization of porosities. The casting methods which can be used here can be listed analogously to the classical classification (Fig. 7.3). They are dealt with in more detail below.

7.3   THE DEVELOPMENT OF NEW CASTING PROCESSES

A notable feature is that the use of pressure has increased with sand- and chill-casting as well, in order to improve feeding and mould-filling per-

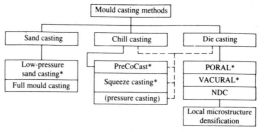

Fig. 7.3   Recent mould-casting methods for aluminium

formance. The pressures used here span the range from less than one bar (for turbulence-free mould filling and the avoidance of gas and oxide inclusions) up to several hundred bar in the solidification phase (to improve feeding performance and minimize shrinkage porosity).

7.3.1 Casting methods using expendable moulds

Casting methods which use lost moulds have been traditionally used mainly for small production runs or to make prototypes. In recent years, however, this type of casting has again become important for larger production series due to machine sand-mould methods with automatic sand metering and compression.

The two new sand-casting methods described below have only led to an increase in economic efficiency of the casting process in one case, however, (7.3.1.2); in the case of low-pressure sand-casting, the main advantage lies in the ability to develop in advance thin-walled components intended for die-casting relatively inexpensively, without having to invest the very high level of expenditure on die-casting tools at the same time. It is very difficult to give generally applicable figures here, as the tooling costs depend on part size and also to a great extent on the complexity of the component, but a cost factor of 10 between a sand-casting pattern installation and a die-casting mould would be a rough approximation. Furthermore, partial modifications to geometry can be made without problems for pattern installations, whereas with die-casting moulds it would be virtually impossible, especially if there are cooling channels and gates.

7.3.1.1 Low-pressure sand-casting

The low-pressure casting method of chill-casting, familiar particularly for wheels, has been adapted by Honsel-Werke to the sand-casting situation. Here the mould is bottom filled, the melt container under the casting mould being subjected to gas pressure (pressure < 1 bar) (Fig. 7.4). Pressure build-up can be programmed as desired with the aid of a microprocessor, so optimal, turbulence-free mould fillings can be set for each component shape. After the mould filling is complete, an excess pressure is maintained in order to post-feed the melt into the casting which is solidifying under volume contraction.

The pressure gives better mould feeding even during the filling phase. This facilitates the manufacture of large surface area, thin-walled components. In addition to small production runs, this method is also useful for making prototypes where the building of an expensive chill during

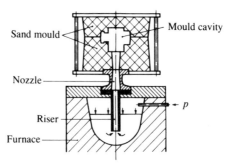

Fig. 7.4   Schematic illustration of low-pressure sand casting installation, from Betz *et al.*

the development and optimization stage would be uneconomical in view of possible mould modifications. Destruction of the comparatively soft sand mould can be countered by applying a facing.

Strength values – in particular with regard to the development of die-cast parts – can be set to be 'die-casting-analogue' by appropriate selection of alloy. In an extreme case this can mean selecting another alloy (type) for strength-related tests on prototype sand-castings, as higher strength values can be achieved with identical alloys due to the faster (better) solidification in the die-casting. The experience of the caster must be used to make this selection.

7.3.1.2   Burnt-pattern casting ('lost foam')
The process patent was lodged by H.F. Shroyer in 1958 (**2**). The principle of the method can be seen in Fig. 7.5. The cavity to be filled with aluminium is embedded in sand as a polystyrene pattern. By joining several sections of polystyrene it is possible to make more complex structures than with one-part castings. As the polystyrene patterns are given a special facing before being embedded in the mould sand, binderless sand can be used, so sand reconditioning is simplified and the amount of used sand is reduced. As there is a shortage of disposal facilities and the costs of disposal are rising, this is surely a factor not to be disregarded.

Separately inserted cores can also be dispensed with, so long as care is taken during manufacture of the polystyrene pattern to ensure that the cavities are adequately filled with sand when the mould sand is shaken in. An additional advantage is the ability to combine several moulds to form a cluster, which can even be 'multi-storey'. The relative number of gates and feeders per casting is thus reduced, and the yield is increased.

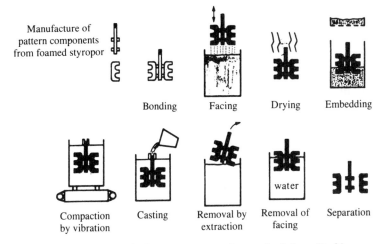

Fig. 7.5   Stages in the burnt-pattern casting method, from Pechine

During the casting process itself, the polystyrene pattern is gassed by the incoming liquid metal. This requires a relatively high casting temperature, entailing the problem of increased gas absorption by the melt, which can cause gas porosity to form in the component.

In a more recent version, pressure is also used in burnt-pattern casting. In the CASTYRAL process, pressure (3 to 10 bar) is used during the solidification phase, which, among other things, considerably reduces the formation of micro-shrink-holes, i.e. shrinkage porosity (**3**).

The process is used in serial production primarily in France and Italy, for components such as inlet manifolds and cylinder heads (**4**). The mechanical properties which can be obtained are equivalent to those found with the conventional sand-casting processes.

7.3.2   Casting methods using permanent moulds

The relatively low casting temperatures for aluminium alloys (660–760°C) and the low ratio of heat content/volume unit permit the almost unrestricted use of the die- and chill-casting permanent mould processes, unlike iron materials. The fast formation of an outer skin and control of attack by the melt on the chill material by means of suitable facings give considerable advantages in the microstructure formation of the casting. These methods have therefore achieved a prominent market position for mass produced parts (Figures for 1989 in the Federal Republic of

Germany: 53 per cent die castings, 33.2 per cent chill-castings, 13.7 per cent sand castings, 0.1 per cent other). While developments in casting methods using expendable moulds are directed more at improving productivity and manufacturability of prototypes, the new casting methods employing permanent moulds generally produce improved characteristics in the component.

7.3.2.1   New chill-casting methods

All the methods listed in Fig. 7.3 under chill-casting have the common feature that mould filling and solidification both take place under pressure.

PreCoCast. Figure 7.6 gives a schematic representation of the process. This is a version of counterpressure casting, in which both the sealed tool chamber and the furnace chamber are filled with air or gas shrouding at up to 15 bar pressure (5). Mould filling is performed as with low-pressure chill-casting, but here it is by reducing the pressure in the tool chamber. While, with low-pressure casting, the maximum refeeding pressure is limited to 1 bar, with the PreCoCast plant a considerably higher differential pressure is possible, which results finally in an improvement in the macro-feeding performance.

In addition, the fact that the mould cavity is subjected to pressure has the advantage that sand cores can be used without risk of being destroyed.

The yield limit can be increased by up to 20 per cent, depending on the sampling location and the alloy (6). Even more marked improve-

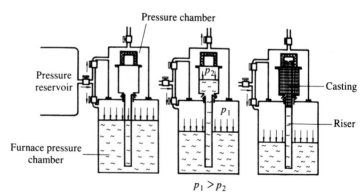

Fig. 7.6   Schematic representation of the PreCoCast method

ments can be achieved with regard to elongation and fatigue strength, which is reasonable in view of the relationship between the pore volume and the process pressure (Fig. 7.7).

Squeeze casting. In principle, 'direct' and 'indirect' squeeze casting are differentiated (7). The former method is also called 'liquid forging' or 'pressure casting' and is illustrated in Fig. 7.8. Firstly mould filling is performed with a precisely metered quantity of melt, then the upper half of the mould is lowered and the mould is closed. Solidification takes place while the pressure is maintained; depending on component size and wall thickness, this may be up to 15 000 bar. The high squeeze pressure ensures good refeeding of the areas which shrink due to volume

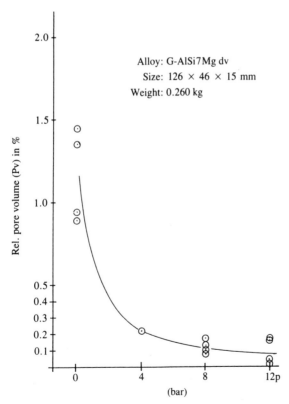

Fig. 7.7 **Relationship between pore volume and process pressure from (6)**

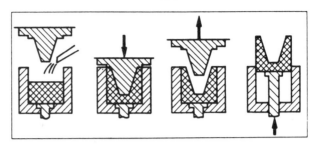

Fig. 7.8   Process stages in 'direct' squeeze casting (7)

contraction during solidification. It also ensures intimate contact between the mould and the metal, which leads to fast cooling and thus to good microstructure formation. The tolerances which can be attained with the casting are within the range for die-casting.

Typical strength values for the alloy AlSi7Mg wa are about 20 per cent higher (R_p) and 10 per cent higher (R_m) than for a conventional chill-casting. The elongation values were improved from 3 per cent to 10 per cent (**8**).

'Indirect' squeeze casting, also known as the UBE HVSC process (**9**), is used mainly in Japan for the manufacture of aluminium wheels. Figure 7.9 shows the principle of the process. Here the melt is not filled directly into the mould, but into a casting chamber which is mounted on a pivoting casting unit and is sealed from below by the casting plunger. After the casting chamber has been filled, the casting unit pivots back, is moved upwards against the fixed half of the mould, and the filling process is started by positioning the casting plunger. The plunger veloc-

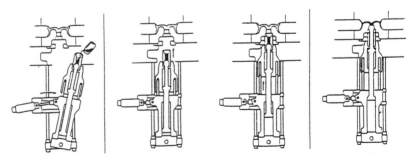

Fig. 7.9   Process stages in 'indirect' squeeze casting, from Ube

ity is selected so as to permit turbulence-free mould filling. In practice this is between 20 and 150 mm/sec. At the end of the mould filling process, a pressure of up to 2000 bar is applied until the completion of solidification, to give castings free of shrink holes.

Comparative analysis of the microstructure of castings manufactured by the 'direct' and 'indirect' methods have shown that microstructure quality is better with the former method. There are, however, some problems associated with the casting process compared with the 'indirect' method. One of these is the possible turbulence during filling of the melt into the mould and the consequent occurrence of oxide skins. Secondly, it is not possible to ensure a constant temperature of the melt, as a relatively long time can elapse before the actual casting process due to the nature of the method.

In general, however, it can be stated that the casting cycle is markedly shorter for squeeze-casting than for chill-casting, and is similar to that for die-casting.

The range of castable alloys does not only include the classical casting alloys – wrought alloys can also be used.

Thixocasting. Another basic way of reducing residual pore-shrink-holes in the casting is to use an 'ssm' technology (ssm = semi-solid metal). The alloys necessary for this differ from the conventional ones not in alloy constituents, but in the microstructure form which gives these alloys the thixotropic properties (**10**). This microstructure is characterized by the fact that the α mixed crystal does not have a dendritic structure, as normal, but has a globulitic structure (Fig. 7.10). This alloy is heated in a temperature range between solidus and liquidus temperature; in practice about 60 per cent remains solid. A material heated in this way behaves like a solid body in the unloaded state. If, however, the piece of metal is placed in the filling chamber of a die-casting machine and subjected to shear stress, the viscosity changes and it behaves like a melt. Figure 7.11 shows the principle of the process.

The casting temperature which is considerably lower than with conventional casting has two advantages. Firstly, the greater proportion of solid phase markedly reduces the shrink porosity due to volume contraction during solidification. Second, the tendency of the metal to absorb gas is reduced, which results in lower gas porosity in the casting. It is this latter argument which demonstrates why this method has been used in serial production, especially in the field of hydraulic elements, as for example brake master cylinders (**11**).

(a)

(b)

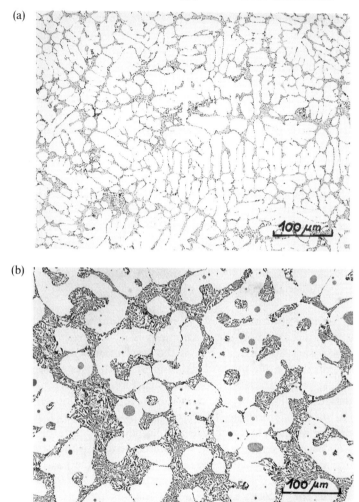

Fig. 7.10   Microstructure comparison of normal casting microstructure with rheo-microstructure (a) normal casting microstructure; (b) rheo microstructure

The limits of the process today are set by the fact that the quantity of metal necessary for large components cannot be heated reliably and uniformly to the precise temperature required for a reproducible casting process. Current development work does, however, give some indication that this problem may soon be solved.

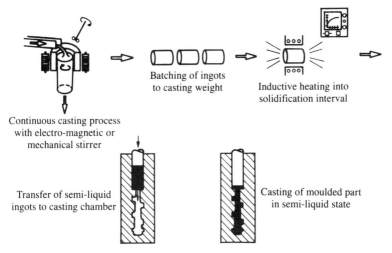

Batching of ingots
to casting weight

Inductive heating into
solidification interval

Continuous casting process
with electro-magnetic or
mechanical stirrer

Transfer of semi-liquid
ingots to casting chamber

Casting of moulded part
in semi-liquid state

Fig. 7.11   **Process stages of thixocasting**

7.3.2.2   Die-casting
While there is still disagreement regarding the process to which squeeze-casting and thixocasting belong, the next two methods described are clearly die-casting methods.

All development work has been directed towards preventing or reducing the advantages associated with the process – the occurrence of pores due to included mould air as well as parting agent and lubricant vapours. This is the main reason why die-cast parts cannot be subjected to heat treatment. With the conventional solution heat treatment temperatures (500–540°C depending on alloy), the strength values of the material are so low that they cannot present any real resistance to the pressure of the included expanding gases, and blow holes form in the component owing to pores near the surface.

Some attempts made to prevent gas inclusions are described below.

PORAL casting. The process variations compared with conventional die-casting are in particular jolt-free operation of the pressure plunger and turbulence-free mould filling (**12**). For each casting, the parameters casting velocity and temperature, cycle time, mould temperature, and quantity of casting metal entering the mould are optimized. This gives castings that are free of gas pores and inclusions, and which may be subjected to heat treatment.

The strength values which can be attained lie in the same range as for appropriately quenched and tempered chill-castings, while elongation values, measured on specimen bars produced from the component, reach approximately 10 per cent with foundry alloy G-AlSi7Mg wa, which is designed for high-grade castings.

VACURAL casting. Another possible way of preventing gas inclusions in a casting is to evacuate the mould cavity and the casting chamber. The companies Müller-Weingarten, Ritter Aluminium GmbH and VAW have carried out a rigorous development programme on this method. The vacuum not only extracts the gases, but also extracts simultaneously the necessary quantity of metal from the liquid metal container under the

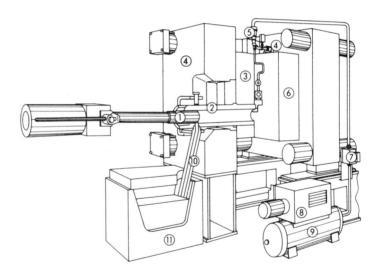

Fig. 7.12   Schematic layout of a VACURAL machine

(1) casting plunger	**(7) valve control**
(2) casting chamber	**(8) vacuum pump**
(3) fixed half of mould	**(9) vacuum tank**
(4) mounting plate	**(10) inlet pipe**
(5) vacuum valve	**(11) holding furnace**
(6) movable half of mould	

casting station via a riser into the casting chamber. Figure 7.12 illustrates the principle of the process.

A particular advantage of the process is that, due to the dual function of the vacuum – namely gas extraction and metal feed in one – defects in the evacuation process become obvious immediately by a lack of metal in the unfinished castings.

Figure 7.13 gives static strength values determined with bar specimens of different alloys. The hardenable alloys AlSi10Mg and AlSi7Mg are particularly worthy of note. Typically they are not used in the die-cast range, as it is not possible to perform 'true' heat treatment (solution annealing, quenching, artificial ageing) in view of the gas pore problem after conventional die-casting. However, their full strength and elongation potential can only be realized through heat treatment. Figure 7.14 gives an example of how specified strength and elongation values can be set, using two diagrams of artificial ageing temperature against time.

Figure 7.15 demonstrates how it is possible to use alloy-specific heat treatment with a high strength level to achieve elongation values which

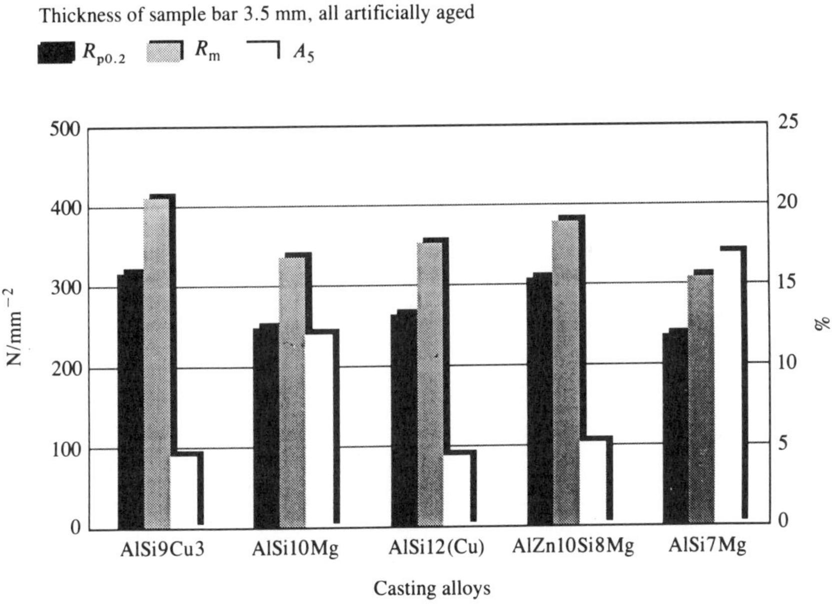

Thickness of sample bar 3.5 mm, all artificially aged

■ $R_{p0.2}$ ▨ R_m ⌐ A_5

Fig. 7.13   Mechanical properties of VACURAL castings

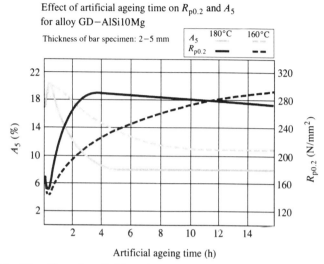

Fig. 7.14   **The effect of artificial ageing temperature against time on mechanical characteristics of VACURAL castings**

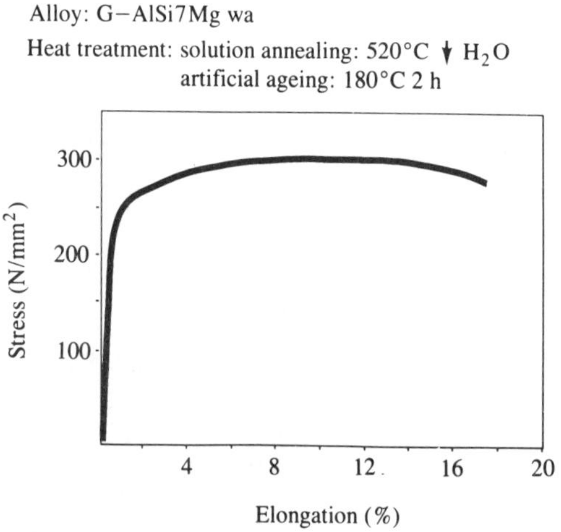

Alloy: G–AlSi7Mg wa

Heat treatment: solution annealing: 520°C ↓ H_2O
                 artificial ageing: 180°C 2 h

Fig. 7.15   **Stress/elongation diagram for a heat treated VACURAL cast bar specimen**

were previously the preserve of wrought materials (sheet, sections, forgings).

There is also a marked improvement in fatigue strength with heat-treated VACURAL castings compared with non-heat-treated specimens (Fig. 7.16). Studies in this field, and on other alloys such as AlSi7Mg will be published shortly.

To reduce shrink porosity which occurs even with this process in areas of greater wall thickness, partial microstructure compaction is used. By this is meant local refeeding under pressure; in practice, movable core pins are used for the ramming. In order to achieve the favourable effect on the microstructure, the ramming pin must be driven into the component at an empirically determined time after mould filling has been completed. If this process starts too late, i.e., if too great a part of the casting has already solidified, the refeeding performance is impaired (13). The situation is illustrated by the porosity measurements in Fig. 7.17.

New Die Cast process (*NDC*). In 1990 Honda presented a new casting process which is important mainly for the economic manufacture of engine blocks (**14**). Whereas, with conventional die-casting methods, it is not possible to use sand cores due to the high metal injection velocities, and 'open-deck' engine concepts must therefore be used instead, with the

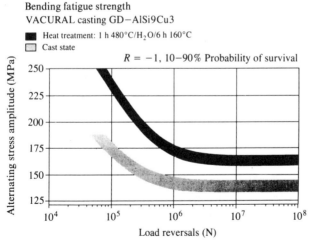

Fig. 7.16 Comparison of the dynamic strength values of VACURAL castings in the heat-treated and the cast state

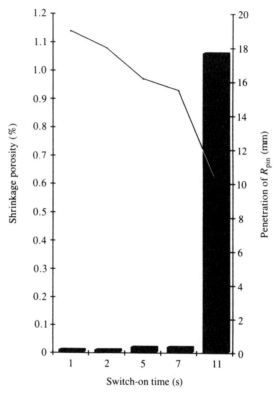

Microstructure compression
AlSi9Cu3

■ Measuring field 1

⁻ Penetration depth of ND pin

Fig. 7.17   The effect of switch-on time on the penetration depth of a ramming pin and the resultant residual porosity

NDC method the flow velocity is reduced to 5 per cent. This permits the use of sand cores and thus the creation of complex cooling channels as well. Due to the lower pressure, the tool can also be made with thinner walls, which facilitates exact measurement and control of tool temperature. This also makes it possible to control the production of a specified microstructure in the casting.

A controlled pressure increase also ensures good refeeding of the metal, as well as up to five times shorter solidification times than with chill-casting processes due to good contact between the metal and the mould. Production cycle times of two minutes are quoted for an engine block, while up to ten minutes must be allowed for such components with chill-casting.

The reduced flow speed during mould filling also prevents gas inclusions due to turbulence, and for this reason these castings can be heat treated.

7.4   NEW ALUMINIUM CASTING MATERIALS

New materials have been developed in recent years on the basis of casting alloys which have been known and largely standardized for decades. They were developed with the following objectives in mind:

- – to increase strength at higher temperatures;
- – to increase strength at room temperature;
- – to increase the elasticity modulus.

This can be achieved, for example, by adding previously unused elements to the alloy (Li) or by casting in other materials (e.g. Al_2O_3 fibres or particles).

7.4.1   Aluminium – lithium casting alloys

The development of AlLi alloys was initiated specifically for the aerospace field. The light element lithium, in addition to providing a significant reduction in specific weight (approximately 10 per cent at 2.5 per cent Li by mass), gives a marked increase in the elasticity modulus and in strength. The base alloy is mainly AlCuMg, which was originally developed to manufacture wrought semi-finished products. Since then, however, we have seen the first applications of precision parts made by precision casting methods for the aircraft industry (**15**).

The manufacture of these alloys is problematic from the technical point of view, as lithium reacts strongly with the atmosphere, so the whole melting and casting process must be performed under gas shielding. This, and the high price of lithium, contribute to an increase in costs, which explains the limitations of possible applications.

The recycling of scrap containing lithium represents another problem. With some wrought alloys, lithium contents of only 2 ppm can have a negative effect on the quality of the semi-finished products (13). For this reason, studies have been performed in which such scrap is added to aluminium recasting alloys. The effects of small lithium contents on the characteristics of the alloys AlSi12Cu and AlSi9Cu3 are described in (16, 17). It was found that a lithium content greater than 0.1 per cent by mass can indeed refine – as desired – the eutectic silicon, but there is a marked effect on the casting porosity, as shown clearly in Fig. 7.18. Although the static strength values are not affected up to a content of 200 ppm lithium (17), it is to be feared that the dynamic strength values will be impaired.

7.4.2   Aluminium composite materials

Here there must be a differention basically between fibre and particle reinforcement; long and short fibres are used in the first type. The fibres are considered to be long if the length is more than 100 times the diameter (18). SiC and Al$_2$O$_3$ fibres are used primarily, while boron and carbon fibres, which have also been studied, can be used only with an additional coating due to their reactivity with the aluminium melt, which makes the already high fibre costs even higher. SiC and Al$_2$O$_3$ particles are also used for particle reinforcement. The particle size is typically 10–15 μm.

While it can be assumed that particle or short-fibre reinforcement gives approximately isotropic material behaviour, reinforcement with long fibres produces clearly anisotropic characteristics. Thus reinforcement with long fibres can permit optimization of the component for the relevant loading. It should be noted here that some of the long fibres are endless fibres. Here the ground fibre is produced by coiling and subsequent treatment with an organic binder. Short fibres are processed either by manufacturing sintered elements, which are then processed into the necessary 'preforms', or by stirring directly into the melt. Blowing in with inert gas, and stirring in, are both standard methods of manufacturing particle-reinforced cast materials.

Processing techniques for casting in/around the 'preforms' include diecasting as well as squeeze-casting and vacuum infiltration. At VAW intensive development is currently underway in the field of fibre-reinforced materials; the VACURAL method is eminently suitable for 'preform' infiltration due to the vacuum present (19).

The mechanical characteristics of a composite material are determined essentially by the volume of fibres it contains. To obtain definite effects,

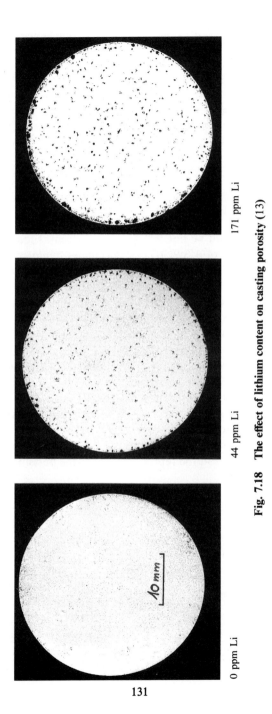

0 ppm Li

44 ppm Li

171 ppm Li

Fig. 7.18   The effect of lithium content on casting porosity (13)

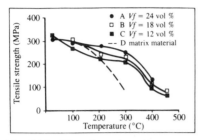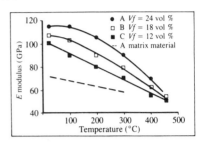

Fig. 7.19   The effect of service temperature on tensile strength and *E* modulus of short-fibre-reinforced AlSi9Cu3 composite materials (20)

the fibre content should be not less than 12 per cent. Figure 7.19 shows the thermal resistance of AlSi9Cu3 cast specimens reinforced with Al_2O_3 short fibres with fibre contents of 12–24 per cent (**20**). Squeeze-casting was the manufacturing method selected. It can be seen clearly that depending on fibre content – there is only a slight decline in strength compared with the matrix material up to about 300°C, but at RT there is no difference. It is different with the elasticity modulus, however; this is markedly higher at room temperature as well.

The above-mentioned anisotropy of the material characteristics can be seen from Fig. 7.20. Whereas, under axial load, i.e. in the direction of the fibres, tensile strength increases markedly with increasing fibre content due to the high strength of the fibre materials, with transverse loading little effect can be detected (**21**).

The success of developing mould castings of fibre-reinforced materials will depend essentially on the reproducibility of the parts. Current devel-

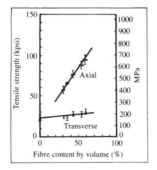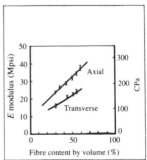

Fig. 7.20   Effect of fibre content on tensile strength and *E* modulus of AlLi2 alloy (21)

opment work therefore is aimed primarily at solving the problem of delamination of the fibre body in the casting phase.

For components which have only very localized high stresses, it is possible to provide local reinforcement in the form of pre- manufactured fibre inserts filled with matrix material. In the current state of knowledge, however, there are certain restrictions regarding the necessary minimum wall thicknesses.

7.5 CONCLUSION

Already, aluminium mould castings occupy a significant position within this family of materials, accounting for more than 80 per cent of the aluminium content of a vehicle. More extensive use of the light metal is prevented by the generally excessively high elongation requirements – particularly in the field of safety-related components – which cannot be fulfilled adequately with conventional casting methods. However, the more recent casting methods, which are now all in use in serial production, will remedy this shortcoming.

8

Solving problems by the use of forgings

H. Lowak and D. Brandt

8.1 INTRODUCTION

Increasing safety requirements and improved comfort, and also the need to meet individual customer requirements, mean that there is ever more equipment in the vehicle and consequently vehicle weight rises and the useful load decreases.

This gives rise to the necessity to save weight in other areas of the vehicle by means of structural measures and by using lightweight materials, such as aluminium alloys, to reduce fuel consumption. As aluminium materials are also eminently recyclable, the needs of environmental protection are met on two counts.

Reducing weight in the chassis area is particularly effective, because the unsprung masses can then be kept small. In addition to reducing petrol consumption, handling is positively influenced. For this reason, most of the aluminium drop-forged parts have been used in this area in the past. Examples include aluminium wheels for passenger cars and trucks, which were originally made exclusively by drop forging with subsequent flow turning (Fig. 8.1).

Suspension arms, longitudinal control arms, and anti-tramp bars are also increasingly being drop forged in highly corrosion-resistant aluminium materials (Fig. 8.2). The same applies to steering shafts and joints of all types (Fig. 8.3). Other examples include ABS pump housings and valve covers as well as the parts for fuel injection systems (Fig. 8.4).

Aluminium articulated shafts with forged joints (Fig. 8.5) are just beginning to be used, while hubs for front and rear axles are so far used only in isolated cases (Fig. 8.6).

8.2 THE FORGING PROCESS

While steels are forged at temperatures of over 1000°C, aluminium materials can be forged at only about 450°C. At these temperatures, aluminium materials can be formed easily, yield stress at low deformation speeds ($\phi = 1$ s^{-1}) is about 50–90 N/mm^2 and is comparable with that of steel.

Fig. 8.1   Forged passenger car wheels

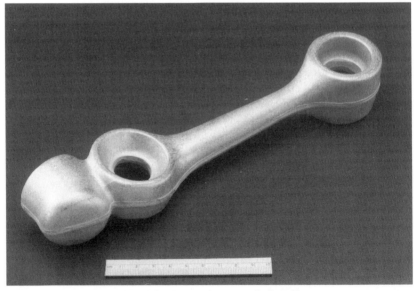

Fig. 8.2   Suspension arm of AlMgSi

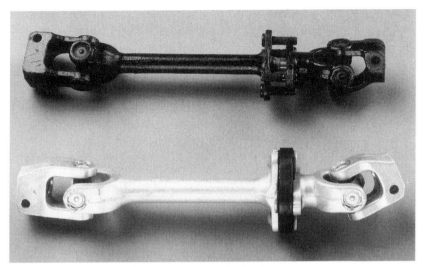

Fig. 8.3 Steering shafts

Fig. 8.4 Drop-forged parts for injection systems and ABS systems

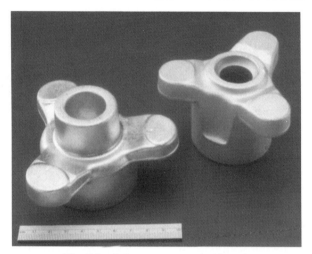

Fig. 8.5   Joint components in AlMgSi

Unlike steels, however, the forging temperature of aluminium materials is not greatly separated from the liquid phase (solidus temperature) (Fig. 8.7). As forming introduces energy and raises the temperature, the forming speed is of decisive importance with aluminium. If

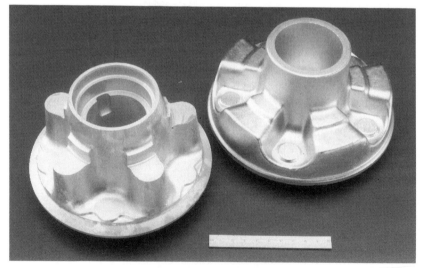

Fig. 8.6   Hubs for passenger cars

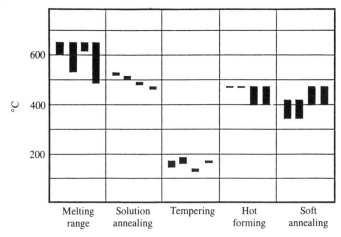

Fig. 8.7   Temperature ranges of some hardenable aluminium alloys

forming is too fast, the temperature rise can be so great locally that the material liquifies and a casting microstructure results on cooling. As it is specifically a wrought microstructure which is required with forgings, the utmost care must be taken to see that the liquid phase is not reached anywhere when forging.

To ensure that this is so, aluminium materials are mainly pressed and not impacted, and so hydraulic presses are used predominantly which have a forming speed considerably lower than that of forging hammers. In borderline cases crank presses and screw presses which have medium forming velocities are also used. The dies are preheated to 350–400°C, so there is virtually no heat flow from the workpiece to the tool. At the same time, the forging process is extended over several minutes in certain circumstances to give the material itself the opportunity of filling the die completely, if the material is viscous.

Aluminium alloys generally do not flow so readily as steel; they therefore have a lower mould-filling capacity. Special measures must be taken to obtain defect-free drop forgings, and this places special demands on the forging expert.

8.3 MATERIALS

The advantages of the forging process come to the fore particularly with hardenable aluminium alloys. Today this means particularly materials of the family AlMgSi, AlCuMg (AlCuSiMn) and AlZnMg (AlZnMgCu).

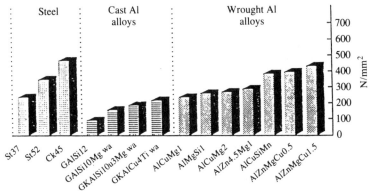

Fig. 8.8 Comparison of yield points (0.2 yield limits) of various aluminium materials with those of steel

Figure 8.8 compares the yield points or 0.2 per cent elongation limits of some steels and standard aluminium casting and forging alloys.

The almost copper-free alloy AlMgSi1 has superior corrosion resistance to all other hardenable aluminium alloys. The material has average strength values and can be welded well, though strength loss in the heat-affected zone must be taken into account. This material is preferred for use in the chassis of passenger cars and commercial vehicles, generally with no surface protection (Fig. 8.9).

Aluminium materials with a higher copper content, such as AlCuMg1 and AlCuMg2, have practically the same 0.2 yield limit as AlMgSi with higher tensile strength and elongation (Fig. 8.10). The materials are not fusion-weldable; surface protection is necessary where there is risk of corrosion. They are used mainly in the cold-worked state and employed where high fatigue strength is required.

The same applies to the alloy AlCuSiMn, which is used in the artificially aged state, is also not fusion-weldable, and requires surface protection in corrosive situations (Fig. 8.11). The material has outstanding strength properties under both static and dynamic loading.

AlZn4.5Mg1 has average strength values, lying between those of AlMgSi and AlCuSiMn. The material can be fusion welded and regains approximately its original strength after welding. For this reason the material is used primarily for highly-stressed welded structures (Fig. 8.12). Subsequent heat treatment is definitely recommended, however, as otherwise there is a risk of stress corrosion cracking. It has average corrosion resistance.

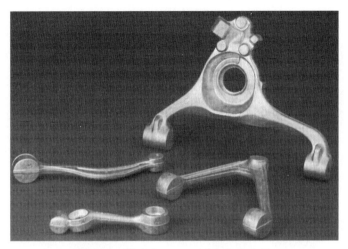

Fig. 8.9   Chassis parts made of AlMgSi1

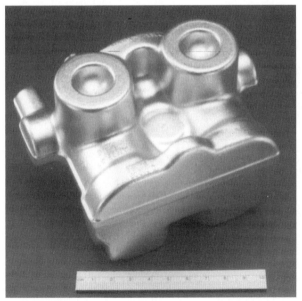

Fig. 8.10   ABS pump housing made of AlCuMg2

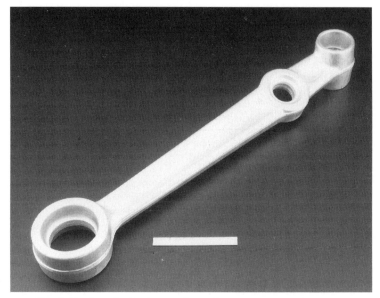

Fig. 8.11   Longitudinal control arm for cars made of AlCuSiMn

Fig. 8.12   Joint made of AlZn4.5Mg1 for propeller shaft

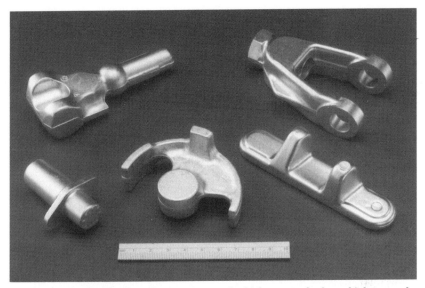

Fig. 8.13 Small drop-forged parts, some of which are made from high-strength aluminium materials

The AlZnMgCu materials achieve the highest strength values of all alloys mentioned (Fig. 8.13). They have the best combination of static and dynamic as well as mechanical fracture properties. They are, in practice, not fusion-weldable and exhibit satisfactory corrosion resistance. These materials are used for the most highly stressed parts owing to their high static and dynamic strength characteristics.

Aluminium forging materials are predominantly quenched and tempered materials. They lose a significant part of their strength at temperatures as low as 100–200°C, depending on material. For this reason, even where operating temperatures are as low as about 100°C it is necessary to decide which aluminium material may be used. The alloy AlCuMgNi has good strength characteristics in this temperature range. It is generally not possible to use aluminium drop forgings at temperatures above 150–200°C.

8.4 DESIGN

An optimal range of good properties can basically only be achieved if the designer is familiar with the characteristics of the manufacturing process and the material when designing the component, utilizes them optimally, and always bears in mind the economic limits. Frequently this is not simple, as he is generally not an expert in this field. Therefore he should bring in a forging expert at the earliest possible stage to discuss with him all aspects which are important for the economic production of the drop forging, namely mass distribution, die partition, minimum wall thicknesses, radii, die inclination, etc. Currently it is possible to make aluminium drop forgings weighing from a few grammes to one tonne. The maximum length of a workpiece is about 5 m, the maximum width about 2 m, and the maximum area transversely to the direction of pressing is about 2 m².

Fibre orientation
The strength, and above all, the fracture elongation and fatigue strength along and transverse to the fibre orientation, are variable. With very highly stressed components it is necessary to guarantee the strength characteristics in the various directions with respect to the fibre orientation.

Figure 8.14 gives one the example of the high-strength aluminium alloy AlZnMgCu1.5 with the properties in the various directions to be guaranteed by the forge. This shows that the characteristics in the direction of the fibre are best. Fibre orientation is therefore very important for highly stressed components. It is particularly decisive for materials which are subjected to fatigue stress (Fig. 8.15).

It must also be borne in mind, however, that each point where a fibre

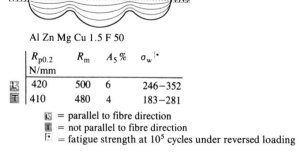

Al Zn Mg Cu 1.5 F 50

	$R_{p0.2}$ N/mm	R_m	A_5 %	σ_w |*
▨	420	500	6	246–352
☰	410	480	4	183–281

▨ = parallel to fibre direction
☰ = not parallel to fibre direction
|* = fatigue strength at 10^5 cycles under reversed loading

Fig. 8.14   Effect of fibre orientation on strength properties of AlZnMgCu1.5

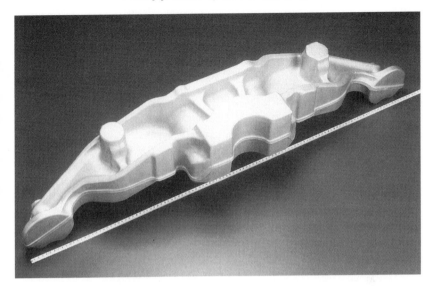

Fig. 8.15   Air bellows mounting for trucks, drop forged from AlMgSi1

leaves the workpiece is an imperfection and thus represents a strength loss in the component at that point. Therefore components machined from the solid do not have the high fatigue strength values of drop forgings (Fig. 8.16).

Only with drop forging is it possible intentionally to match the fibre orientation to the main lines of force flow using appropriate measures, and thus to impart the best possible strength properties to the workpiece; this fact generally means that drop forgings are automatically used for highly stressed safety components (Fig. 8.17).

Machined   Drop forged

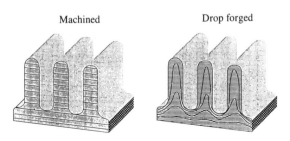

Fig. 8.16   Fibre orientation in machined and forged cross-sections

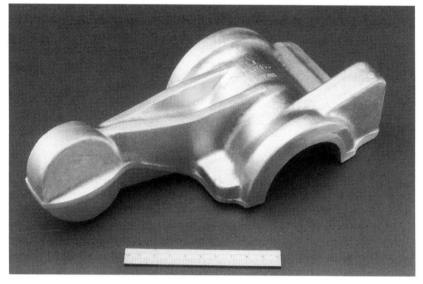

Fig. 8.17   Drop-forged bearing housing

8.5 DIMENSIONING AND RATING

Components in the vehicle structure are generally subjected to cyclic loading, so, depending on area of application, they must be designed for fatigue strength or service strength. In the region of the drivetrain (engine and transmission), design must be for fatigue strength due to the very high number of load reversals (each revolution of the engine is a load cycle), while other areas are designed for service strength, that is, their dimensions are selected on the basis of a load pattern. In addition to strength, stiffness of the individual components is important for a vehicle, which is a system capable of vibration, i.e. the elasticity modulus is also important as a graduator for elastic deformation. The elasticity modulus of aluminium is only one third that of steel. If necessary, this difference can easily be counterbalanced by design measures. In these cases, the weight saving achieved by using aluminium is generally reduced by about 50 per cent.

In the case of components subjected to pulsating loads, which must be dimensioned very close to the limit for reasons of weight-saving, i.e., which are highly stressed, particular attention must be paid to achieving a design which is appropriate to the material. Wrought aluminium alloys

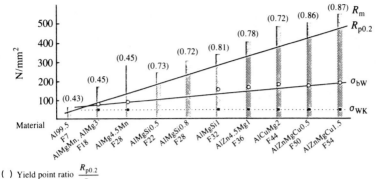

( ) Yield point ratio $\dfrac{R_{p0.2}}{R_m}$

ꝑ Fatigue strength under reversed stresses σ_{bW}
▓ Notch fatigue strength under reversed stresses σ_{WK}

Fig. 8.18 **Comparison of 0.2 yield point with bending fatigue strength and notch fatigue strength of some aluminium wrought materials (R_m = maximum tensile strength; $R_{p0.2}$ = yield strength)**

used in automotive construction are basically quenched and tempered materials, which, like highly quenched and tempered steel, are sensitive to notches. So, particularly with components subjected to fatigue stress, the higher static strength can only be converted into higher fatigue strength if the components are designed with few notches (Fig. 8.18). Transitions of cross-section must therefore be designed with radii which are as large as possible (Fig. 8.19). This applies likewise to components subjected to service stress. It is true that, in this case, higher static strength gives a certain advantage in dimensioning even with severely notched components.

Chassis components of vehicles are as a rule safety parts which must

Risk from notches – not recommended

Few notches – better

Fig. 8.19 Avoidance of notches

meet special requirements. On the one hand, one has to ensure that they can withstand the stresses occurring during normal use of a vehicle over the whole life of the vehicle without failing, thus fulfilling their function safely. This is ensured by designing for service strength. At the same time, however, it is necessary to ensure that if the vehicle is misused or is involved in accident-like situations, they fulfil their function in such a way that there is no danger of death or injury for the vehicle user or other road users.

Chassis parts with such a 'good-natured' loading characteristic are plastically deformed when overloaded. Such a permanent deformation of a special component imposed from outside prevents damage to other, more sensitive components behind it which are less accessible for examination. So, for instance, in a low-speed offset impact with an obstacle, the steering gear remains intact, while the transverse suspension arm is permanently deformed. The change in the handling of the vehicle due to this deformation makes the driver aware of the damage caused. The damaged parts can thus be replaced without anyone being put at risk.

Special demands are therefore placed on the toughness and deformation behaviour of such components, which can be met particularly well by drop-forged parts. Figures 8.20 and 8.21 compare the static strength and fatigue strength properties of two aluminium alloys frequently used in vehicles which illustrate this fact. These are the wrought alloy AlMgSi1, F31, from which most aluminium forgings in the chassis area are made, and the artificially aged casting alloy GK-AlSi7Mg, which is used for wheels, for example. The strength values given were determined from specimens taken from manufactured components.

A wrought alloy has advantages of strength and toughness under static load and thus has greater safety reserves with regard to impact stress. A comparison of elongation stress-number curves, which indicate the behaviour of the material under repeated stress, also shows wrought alloys in a favourable light. In addition to considerably better fatigue strength, a wrought alloy has a considerably higher acceptable elongation amplitude than a cast alloy in the short-term strength range, i.e., it is able to dissipate excessive stresses by means of its very good plasticity. From this it can be concluded that a wrought alloy also has the better mechanical fracture characteristics.

Fatigue cracks which occur in wrought alloys propagate more slowly than in cast alloys, and the critical crack length, which leads to final failure of the component, is greater than for a cast alloy. A comparison

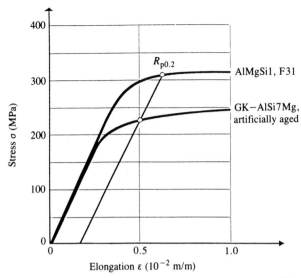

Fig. 8.20   Stress–strain curves for AlMgSi1, F32 and GK-AlSi7Mg

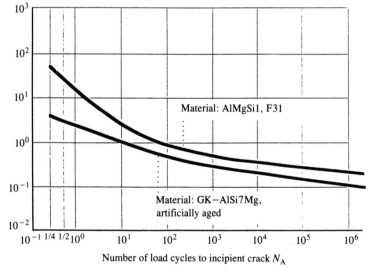

Fig. 8.21   ε–N fatigue curves for AlMgSi1, F31 and GK-AlSi7Mg

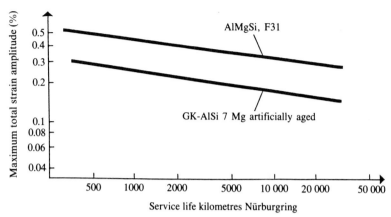

Fig. 8.22   Service life curves under simulated operating stresses for AlMgSi1, F31 and GK-AlSi7Mg

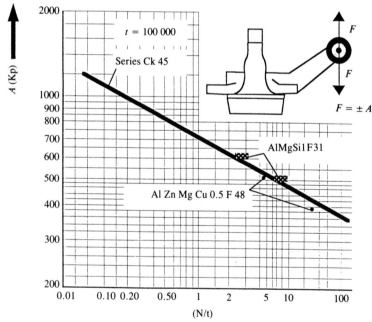

Fig. 8.23   Aluminium pivot bearing: test of control arm (by Ostermann)

of service life curves in Fig. 8.22 shows how the two materials behave under the actual stresses occurring during operation.

One example which shows that even very complicated components subjected to complex stresses can make the transition from steel to aluminium is the pivot bearing shown in Fig. 8.23. In the transition from Ck 45 to AlMgSi1, F31, optimal dimensioning with the same service life allowed the weight to be reduced by half while ensuring the required deformation characteristics were produced. This example demonstrates that there are virtually no technical limits for making forgings from aluminium, so long as there is sufficient installation space. The limits are set by the economic efficiency, which will be dealt with in more detail below.

8.6 ECONOMIC EFFICIENCY

Minimum production quantities
In the case of drop forgings, the minimum production quantity depends greatly on the size of the workpiece. The smaller the workpiece, the larger is the minimum production quantity for the drop forging to be worthwhile. With large workpieces, lots of 100 may be worthwhile in certain circumstances, while with small components 1000 is often too small a quantity. Here the ratio of component value to work involved is important.

Large-scale production
In modern large-scale production – this assumption is generally true for the large quantities required in the automotive industry – manufacturing factors must be taken into account to a greater extent when designing the components, than with small- and medium-sized production runs. In certain circumstances slight concessions on component weight can considerably improve the economics of production.

So, for example, turning the land through 90° for a passenger car transverse suspension arm (Fig. 8.24) entailed a slight weight increase but gave noticeable savings in manufacture. This cost-saving measure made the drop-forged aluminium suspension arm competitive with the forged steel part, and it is used in production today due to its weight advantage.

Large-scale production is generally performed on presses which are largely automated and operated by robots. Such automated production requires lot sizes of at least 30 000 parts.

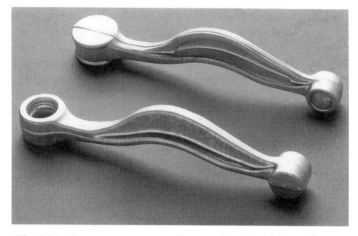

Fig. 8.24   Car transverse suspension arm drop forged from AlMgSi1

Tooling costs

The die costs for aluminium and steel forgings are about the same. It is true that the die surface must meet greater demands in the case of aluminium, so die maintenance is more expensive than for steel.

On the one hand this is associated with the lower mould-filling capacity of aluminium already mentioned, and on the other, the surface of the aluminium forging is generally not subjected to further treatment except for slight etching, while with steel, removing the layer of scale entails surface treatment.

Tooling costs can be influenced greatly by the mass distribution of the part being made. The more unfavourable this mass distribution is, the more forging steps with associated dies are necessary. Naturally, tooling costs are also dependent on the size and shape of the drop forging.

It is virtually impossible to make generally applicable statements on tool life, because the shape of the component is a decisive factor here. It is roughly comparable with that for steel. As already mentioned, die maintenance is more expensive than for steel.

Unit prices

It is not possible to give general guidelines as to price. However, the following can be assumed as a very rough estimate.

With smaller production runs and conventional manufacturing methods, aluminium drop forgings cost about twice as much as corresponding aluminium castings. Compared with steel drop forgings, the price of aluminium drop forgings is approximately 50 per cent higher. With large-scale production on fully automatic forging presses, the additional forging and heat treatment can be kept within acceptable limits, particularly if the special requirements of aluminium have already been taken into account during design of the components. Due to the excellent machinability of aluminium, expense can be saved compared with steel in the final treatment. In the case of components which are not so material-intensive, as is usually the case with chassis parts, the additional costs therefore lie within a range of magnitude which is clearly counterbalanced by the advantages offered.

Basically, it can be stated that:

(a) An aluminium forging is more expensive than a corresponding steel part due to the price of the material.

(b) The economics of the component cannot be judged solely on the basis of the price of the semi-finished product.

Whether the higher price of an aluminium part is justified will only be determined when the vehicle is in operation, namely if, for example, the payload can be increased correspondingly, or if it is possible to fit auxiliary equipment in view of the weight saving, or even if the vehicle achieves the correct performance characteristics only as a result of this lower weight, or petrol consumption is improved.

8.7 CONCLUSIONS AND PROSPECTS

Forging is the only process with which it is possible consciously to achieve a parallel relationship between fibre orientation and the main force flow lines by using specific measures. This means that aluminium drop forgings can be used wherever good fatigue strength and ductility in the material are of decisive importance, that is, particularly in highly stressed safety components.

As aluminium forgings are made of wrought materials which are completely free of shrink holes and pores, they are liquid-, air-, and vacuum-tight and thus eminently suitable for all components subject to pressure.

The low specific weight combined with good strength and toughness and excellent corrosion characteristics as well as almost 100 per cent recyclability make aluminium wrought alloys an ideal material for vehicle construction. If the characteristics of aluminium are fully utilized in a new design, and the special production factors are taken into account, optimal solutions can be found with regard to the economics of the material as well.

The aluminium industry is making great efforts to develop new materials for applications which were previously not accessible to this group of materials.

In this connection we can think mainly of aluminium materials manufactured using powder metallurgy and incorporating fibre reinforcement, which can provide special material characteristics, such as improved thermal resistance (service range up to 350°C), high static strength (over 700 N/mm²), and high fatigue strength with low notch sensitivity.

A new group of materials, which are used already in the aerospace industry, are aluminium alloys containing lithium, which have a 10 per cent lower specific weight and a 10 per cent higher *E* modulus. These new materials are still largely in the development stage. Whether they will find applications in the motor industry despite their currently high cost remains to be seen.

The potential of today's conventional aluminium forgings is by no means fully exhausted yet. So designers in the motor industry still have numerous opportunities for using aluminium, and the aluminium forgings industry is ready to be an active partner in developing and implementing them.

Principles of drawing aluminium body parts

F. Ostermann

9.1 INTRODUCTION

All the basic processes, tool-making methods and machinery which have been developed for making body panels from steel sheet can be used for drawing body parts made of aluminium sheet. However, the differences between the two groups of materials are so great (1) that, in general, and without adaptation to the specific characteristics of the material, tools and billet blanks cannot be used simultaneously for the two materials with the same success.

The forming characteristics of aluminium body sheet materials are therefore dealt with below.

9.2 STRESS APPLIED TO THE SHEET MATERIAL BY THE DRAWING PROCESS

Drawing body sheet is generally a very complex forming process, in which different forming processes take place simultaneously, next to one another, or one after the other, and subject the material to different stresses. The specific forming processes include:

– deep drawing;
– stretch forming;
– hole widening;
– beading;
– bending;
– seaming.

During these processes the forming zones experience different strain and elongation states (deformation states). The forming limits of the material are dependent on the strain and elongation states pertaining and on their variation during the course of the forming process ('deformation history'). The forming limit in sheet forming is generally determined by the start of necking down and in bending and seaming by incipient microcrack formation.

Figure 9.1 shows a forming limit diagram by Nakazima and other authors (2), in which the effects of the deformation state and the deformation history during sheet forming can be seen. This diagram contains the main deformations ϕ_1 and ϕ_2 occurring in the sheet plane, which can be determined for example, with line networks or graduated circles applied previously to the unformed sheet. The deformation states along the (dashed) lines 1, 2 and 3 relate to ideal forming processes

deep drawing ($\phi_1 = -\phi_2$),
uniaxial tensile test ($\phi_1 = -2\phi_2$) and
stretch drawing ($\phi_1 = \phi_2$).

The curves given for the start of necking show clearly that the formability limits of a material can be indicated only for ideal forming pro-

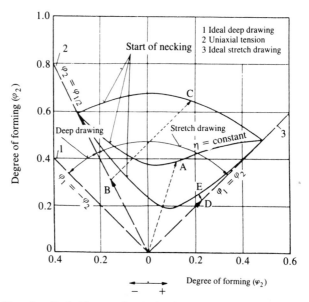

Fig. 9.1   Forming limit diagram by Nakazima and other authors (2). Effect of deformation path on the position and shape of the limit curve for the start of necking

cesses. These limits are not a reliable indication of the values for real drawing processes.

In developing a new body sheet part or in testing a new sheet material for an existing body part, however, the deformation analysis is a considerable help. By following the deformation path in critical areas where there is risk of cracking, it is possible to devise and implement measures to achieve reliable manufacture.

Evaluating the malleability of a material can therefore currently only be performed with the aid of several forming parameters. These include data from the uniaxial tensile test, the maximum deep drawability $\beta_{0\max}$, the stretch drawability, which is determined with the Erichsen cupping index I_E or in the hydraulic cupping or bulge test, and the minimum bending or seam radius. Other details, such as the tribological characteristics of the sheet surface, the change in the surface structure due to the forming process, and the behaviour of the material in contact with the tool surface are also necessary in order to obtain satisfactory drawing. These are dealt with in the following sections.

9.3 FORMABILITY CHARACTERISTICS IN THE TENSILE TEST

Characterizing the forming performance of a sheet material with the aid of the uniaxial tensile test is important in that the tensile test provides values which are largely independent of geometry, sheet gauge and friction effects. At the same time, the tensile test can be used to determine, simply and reliably, the anisotropy of the flow characteristics of the sheet in the sheet plane by appropriate selection of the position of the specimen.

The conventional measurements obtained in the tensile test, namely 0.2 yield point ($R_{p0.2}$), tensile strength (R_m) and fracture elongation (A_5, A_{10}, or A_{80}), are indeed important for quality assurance and as rating values, but for assessing forming characteristics and forming behaviour, other information from the tensile test is more useful. This information relates to the flow curve ($k_f = f(\phi)$), the uniform elongation (A_{gl}), the hardening capacity (n) and the vertical anisotropy (r), which are described in greater detail below.

(a) Flow curve
The flow curve is obtained in the tensile test with a tensile specimen, by relating the tensile force F to the instantaneous specimen cross-section

S ($k_f = F/S$) and plotting it over the degree of forming $\phi = \ln l/l_o$ (l = instantaneous measured length, l_o = initial length):

$$k_f = k_f(\phi) \tag{1}$$

It is usual to describe the flow curve by means of the so-called 'Ludwik equation':

$$k_f(\phi) = C \cdot \phi^n \tag{2}$$

Here C is a material constant and n the hardening exponent.

With the aid of the 0.2 yield limit $R_{p0.2}$ ($k_f = R_{p0.2}$; $\phi = \ln (1 + 0.002)$ and of the hardening exponent n, the flow curve can thus be calculated approximately with equation (2). The Tresca and von Mises flow conditions can be used for conversion to multi-axial stress states.

For aluminium bodysheet materials, the necessary information on $R_{p0.2}$ and n is given in Table 1.5 of Chapter 1.

Determining the flow curve with the uniaxial tensile test is restricted to the range of uniform elongation, A_{gl}. Higher forming levels can be achieved with sheet materials such as the hydraulic bulge test. Flow curves for the body materials AlMg3, AlMg5Mn and AlMg0.4Si1.2, determined with the instrumented bulge test, may be found in (**3**).

(b)   Hardening exponent n

Aluminium body sheet characteristically has a high hardening capacity, which is expressed in high n values, which lie in the range between $n = 0.25$ and $n = 0.35$. Alloys with a high magnesium content, for example AlMg5Mn, are superior to all other body sheet materials – even high-grade steel sheet – in this. A high hardening capacity has a positive effect on the forming behaviour, especially when a high flow resistance is required in areas with a high elongation load under stretch drawing conditions, because the deformation distribution thereby becomes more uniform and less-hardened sheet areas used in forming.

The values of n given in various literature references can, however, only be used with some reservation for a more quantitative assessment of formability, as tests with new temsile testing machines with computer-based evaluation methods have indicated a marked dependence on the degree of forming ϕ. At the present time there is no standard test method for determining the n value of aluminium sheet.

(c)   Uniform elongation, fracture elongation

Aluminium body sheet exhibits lower fracture elongation than deep drawing grades of steel. Figure 9.2 shows schematically stress-elongation

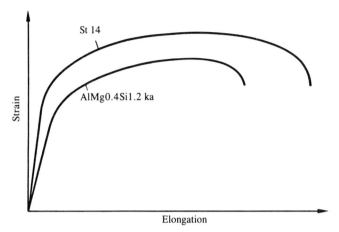

Fig. 9.2 Schematic representation of stress-elongation curves for St14 and AlMg0.4Si1.2

curves for steel 14 and AlMg0.4Si1.2. It can be seen that steel has a greater necking elongation than the aluminium alloy. The necking elongation is the difference between the fracture elongation and the uniform elongation and starts when the maximum acceptable stress, i.e. the tensile strength, is exceeded. So, unlike steel, when drawing aluminium body parts deformation should not go beyond uniform elongation, otherwise there is a risk of cracking.

Blaich (4) measured the elongation distribution over the measured length of tensile specimens of various body materials stressed to fracture (see Fig. 9.3). The greater necking range of RRSt1403 compared with AlMg5(Mn), for example, is noticeable. On the other hand, it also shows clearly the more uniform deformation distribution of the greatly hardened AlMg5(Mn) alloy.

Uniform elongation has not normally been one of the parameters obtained in the tensile test, as its determination is subject to uncertainty in assessment of the test. It can however be calculated approximately from the values for fracture elongation A_5 and A_{10} using equation (3):

$$A_{gl} = 2A_{10} - A_5 (\%) \tag{3}$$

A calculation of uniform elongation using equation (3) assumes, however, that the necking elongation is in the range of the fracture elongation A_5 (5), which is not so in the case of RRSt1403 in Fig. 9.3.

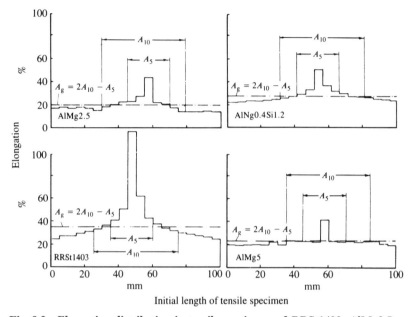

Initial length of tensile specimen

Fig. 9.3   Elongation distribution in tensile specimens of RRSt1403, AlMg2.5 w, AlMg5(Mn) w and AlMg0.4Si1.2 ka [4]

Typical uniform and fracture elongation values of aluminium body-sheet materials can be found in Table 1.5 of Chapter 1.

(d)   Plastic anisotropy, r value

In the case of a tensile specimen from an isotropic sheet material, which is elongated by the amount ϕ_1 in the axis of the bar, the deformation in the specimen width ϕ_b and in the specimen thickness ϕ_s are the same:

$$\phi_b = \phi_s \tag{4}$$

For reasons of volume constancy

$$\phi_1 + \phi_b + \phi_s = 0 \tag{5}$$

and

$$\phi_1 = -2 \cdot \phi_b = -2 \cdot \phi_s \tag{6}$$

If, however, the condition of equation (4) is not met, i.e. the deformations in the specimen width and specimen thickness are different, the material is anisotropic. The ratio ϕ_b/ϕ_s is called perpendicular anisotropy r

$$r = \phi_b/\phi_s \tag{7}$$

If deformation in the specimen width is greater than in the specimen thickness ($r > 1$), the deformation resistance is correspondingly greater in the specimen thickness than in the specimen width. Such a relationship has a favourable effect on the deep drawability and especially on the stretch drawability of a sheet material.

The r value is dependent on the position of the specimen relative to roll direction. It is therefore normal to determine the r value in the directions 0°, 45° and 90° to the roll direction, and to calculate from them the mean perpendicular anisotropy as follows:

$$r_m = (r_{0°} + 2r_{45°} + r_{90°})/4 \tag{8}$$

Figure 9.4 gives typical r values for three aluminium body sheet alloys in the various orientations to the roll direction (**4**).

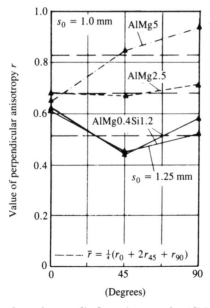

Fig. 9.4 **Typical r values of perpendicular anisotropy for AlMg2.5 w, AlMg5(Mn) w and AlMg0.4Si1.2 ka, measured at 0°, 45° and 90° to the roll direction**

Plastic anisotropy is determined by the rolling and annealing textures of the sheet, which in turn are dependent on the alloy composition, the manufacturing parameters, and above all on the slip systems in the crystal lattice of the base metal. Ferritic sheet has average r values between 1 and 2, whereas aluminium body sheet materials have average r values of between 0.5 and 1, when in the annealed state. This can be regarded as one of the causes of the lower stretch drawability of sheet made of aluminium alloys compared with steel sheet.

Table 1.5 in Chapter 1 gives a summary of typical r_m values for various aluminium body sheet materials.

9.4 DEEP DRAWING AND STRETCH FORMING PROPERTIES

Deep drawing, as per DIN 8584, is the forming of a presheared sheet into a hollow body open on one side without intentional alteration of the sheet thickness. The deep drawing property of a material is described by the limiting drawing ratio β_{0max}. It is the ratio of the largest circular blank diameter D that can just be shallow formed without cracking to the punch diameter d.

$$\beta_{0max} = D/d \tag{9}$$

The limiting drawing ratio increases with increasing uniform elongation and rising r_m value (6). On the other hand, the friction ratios under the blankholder are at least as important for the deep drawing result as differences in uniform elongation and r_m values between deep drawing grades of various types of material and between steel and aluminium. Basically, one can expect good deep drawing properties in aluminium body sheet materials, as they all have a limiting drawing ratio β_{0max} of about 2.0–2.1. For comparison, deep drawing grades of steel have β_{0max} values between 2.0 and 2.3 (6). While the sheet thickness remains approximately constant with pure deep drawing, in stretch forming the forming takes place at the expense of the sheet thickness and therefore leads to an enlargement of the sheet surface under the biaxial tensile stress. The characteristic value usually determined for stretch formability is Erichsen cupping I_E (mm), which is standardized in DIN 50101 and DIN 50102. There is a certain uncertainty in the assessment of comparative figures for Erichsen cupping due to the influence of friction between the punch head and the sheet. This frictional influence is avoided in the hydraulic cupping test, but due to the greater complexity of the test it is not so frequently used as the Erichsen cupping test.

Another difficulty in comparing different cupping values is their dependence on the initial sheet thickness, so only values obtained with the same sheet thickness permit assessment of stretch formability. The dependence of the cupping value on sheet thickness does not appear to be linear over the sheet gauge range 0.2–2.0 mm (7). For soft-annealed aluminium materials, however, one can expect an approximately 10 per cent higher cupping value for an increase in sheet thickness of 0.2–0.3 mm.

By correlating cupping values obtained in Erichsen and hydraulic cupping tests on numerous sheet materials with suitable characteristic values in the tensile test, it has been possible to define a so-called 'stretch formability value' S, which can be calculated simply using the data from the tensile test as follows (see Fig. 9.5) (8):

$$S = R_m/R_{p0.2} + n_m + 2r_{min} \tag{10}$$

Here

R_m = tensile strength
$R_{p0.2}$ = 0.2 yield point
n_m = average hardening exponent
   = $(n_{0°} + 2n_{45°} + n_{90°})/4$
r_{min} = smallest value of perpendicular anisotropy

The apparent absence of a ductility or elongation value in equation (10) can be explained by the fact that when using the Ludwik equation (see equation (2)), the hardening exponent n can be equated with the uniform elongation ϕ_g.

If the different materials are analysed with regard to hardening influence ($R_m/R_{p0.2}$ and n_m) and texture influence (r_{min}) on the value S, it is found that, irrespective of the magnitude of the S value for aluminium and austenitic special steel, the stretch formability depends 65 per cent on the hardening properties and 35 per cent on the texture properties. With ferritic steel materials, however, the texture component is 55 per cent and the hardening component is 45 per cent. These differences are doubtless due to the different crystal structures (fcc or bcc) of the two groups of materials (9).

In accordance with equation (10), changes in the minimum value of perpendicular anisotropy, r_{min}, can be of considerable importance for

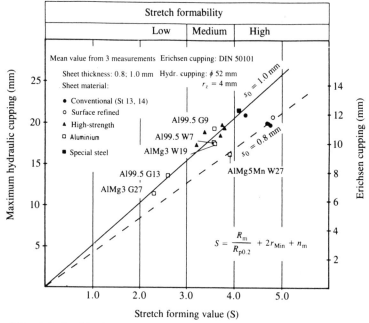

Fig. 9.5   Relationship between the stretch forming value S and cupping values from the Erichsen test and the hydraulic cupping test (8)

stretch formability due to the factor 2. A uniform r value distribution in the sheet plane should therefore be advantageous for forming behaviour.

In conclusion, the above presentation of the deep drawing and stretch forming properties of aluminium materials leads to the following basic factors involved in drawing body parts made of aluminium compared with those made of steel. In view of the comparatively good deep drawing properties and the poorer, texture-related stretch forming properties, the drawing stages of difficult parts should be arranged so that deep drawing is used first as the main forming process and that the stretch forming process is only initiated at a later stage. Such a procedure provides a favourable sequence for the deformation path, as indicated ideally for example in the forming limit diagram in Fig. 9.1 by the sequence O → B → C.

9.5   TRIBOLOGY IN THE FORMING PROCESS

In addition to the forming properties of the material, the tribological conditions in the contact zones between the sheet surface and the tool surface play an important part in determining the procedural limits of the forming process. The friction in the various contact zones affects the flow of the material in the tool and is used deliberately to control the forming process.

The friction zones in deep drawing and in body part drawing are illustrated schematically in Fig. 9.6 **(10)**, **(11)**. The demands made on the friction situation in these friction zones can vary greatly depending on the type of part being drawn and the forming procedure. In deep drawing, low friction is required under the blankholder (zone 1) and at the drawing die curvature (zone 2), in order to reduce drawing forces. At the punch edge (zone 3), friction needs to be as high as possible, so that high forces are introduced into the cup wall at the transition zone from punch to cup wall. If special areas have to be drawn out by stretch forming in the base of the drawn part, low friction values are desirable at the punch face (zone 4). To control material flow in the case of irregular drawn parts, such as, for example, body parts, higher friction may be necessary in certain parts of the blankholder, which can be achieved with locally higher surface pressure or with braking bulges (drawing beads).

The tribological system as a whole consists of the sheet surface, the tool surface and the lubricant. The lubricant prevents abrasion and wear of the tool and workpiece surfaces, and, particularly in the case of drawn aluminium parts, prevents adhesion at the tool surface. At the same time, it is necessary to keep the use of lubricant as low as possible. The capacity of the sheet surface to absorb lubricant and thus the precision surface structure of the sheet are correspondingly important.

As already indicated in Chapter 1, the standard rolled surface is the so-called 'mill-finish' surface with relatively low roughness coefficients. It

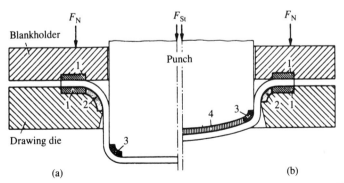

Fig. 9.6   Friction zones in deep drawing and in body part drawing (10, 11)

is produced with tangentially ground rolls and thus exhibits a directional roughness, which produces different tribological behaviour parallel to and transverse to the roll direction. This anisotropy of the friction performance of body sheet materials AlMg2.5, AlMg5(Mn) and AlMg0.4Si1.2 is shown in Fig. 9.7. The results were obtained in a strip drawing test without reversion (12). It can be seen that cold welding occurs at relatively low blankholder pressures in the selected conditions, and is dependent on the orientation of the strip drawing direction with respect to the roll direction. The dependence of the drawing forces on orientation to the roll direction with a given blankholder pressure is also evident.

For this reason, attempts have been made in recent years, as with high-grade dressed sheet steel, to develop special surfaces with non-directional roughness. The test data in Fig. 9.8 are an example of the tribological behaviour of various surface preparations of the body material AlMg5Mn (11).

With regard to a tendency to seize, the laser-textured and eroded surfaces behave optimally. The smooth-blasted surface also exhibits markedly better behaviour than the ground surfaces in the given conditions. The favourable tribological characteristics of these surfaces have an advantageous effect on deformation distribution, as Fig. 9.9 shows (11).

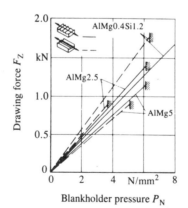

Fig. 9.7   **The influence of the sheet material (AlMg2.5, AlMg5(Mn), and AlMg0.4Si1.2) and of the anisotropy of the friction behaviour of mill-finish surfaces on drawing forces and the occurrence of cold welding (12). Strip drawing test without reversion, drawing rate 2 mm/s, drawing distance 100 mm, lubricant Oest Platinol V711/80.**

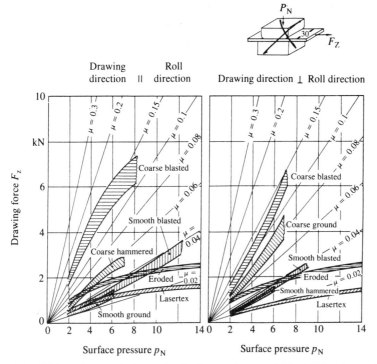

Fig. 9.8 **Sliding characteristics and adhesion tendency of AlMg5Mn sheet with different surface preparations in the strip drawing test (11).** *Left:* **drawing direction parallel to roll direction.** *Right:* **drawing direction perpendicular to roll direction. Tool material GG25CrMo, lubricant M100, drawing rate 100 mm/s, drawing distance 100 mm.**

The second friction partner, the tool surface, also makes an important contribution to the tribological situation. Basically the same tool materials are used for drawing aluminium body parts as for manufacturing steel bodies, for example cast iron GG26, GG25CrMo and tool steel inserts for drawing edges, drawing beads and cutting edges. To prevent the occurrence of adhesion, the roughness of the tool surface in critical contact zones should meet the following requirements **(12)**, **(13)**:

$$R_z \leq 1 \ \mu\text{m}$$

$$\lambda_p \geq 0.46$$

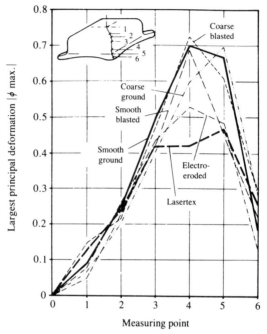

Fig. 9.9   Influence of various surface preparations on the largest principal deformation in the formation of a rectangular cup in base material AlMg5Mn (11). Drawing gap u_z/s_o = 1.4, drawing edge radius r_z = 10 mm, blankholder pressure p_N = 8 N/mm², drawing rate 100 mm/s, lubricant M100, quantity of lubricant 4.5 g/m².

where

R_z = average peak-to-valley depth

λ_p = degree of profile emptiness = R_p/R_t

R_p = peak to mean line height

R_t = peak to valley height

The adhesion tendency and the friction coefficient can also be particularly effectively influenced by surface treatment of the tool, as shown in Fig. 9.10 (**12**). The type of surface treatment which is particularly suitable for a specific case depends on the technical and economic parameters, such as the size of the tool.

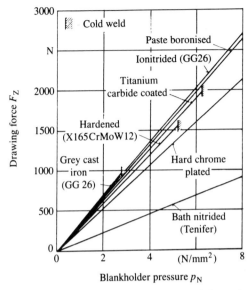

Fig. 9.10   **Influence of material and surface treatment of the tool on drawing force and the incidence of cold welding (12). Strip drawing test without reversion with AlMg0.4Si1.2 ka parallel to the roll direction, lubricant Oest Platinol V711/80, drawing rate 2 mm/s, drawing distance 100 mm.**

Table 9.1   Recommendations for the design of drawing tools for aluminium body sheet with regard to drawing and cutting gaps, drawing radii and drawing beads

Drawing gap	u_z/s_o	1.1
Drawing edge radius	r_z/s_o	
at straight edges	r_z/s_o	5–7
in corners	r_z/s_o	6–10
Punch radius	r_{st}/s_o	6–12
corners	r_{st}/s_o	12
for reinforcing struts	r_{st}/s_o	10–15
Drawing beads		
width	$2r_B/s_o$	12
height	h/s_o	6–10
Cutting gap u_s/s_o		0.06–0.08

s_o = initial sheet thickness

The appropriate design of drawing and cutting tools for aluminium also involves the data for drawing and cutting gaps and tool radii, as indicated in Table 9.1 (from Wolff (**14**)). The cutting edges of blanking tools should be greased to prevent pick-up and the transfer of abraded particles to the sheet bar and into the drawing tool.

The successful drawing of aluminium body parts also depends decisively on the choice of a suitable lubricant and its application to the sheet bar and the tool. A detailed presentation of this problem area can be found in (**15**). Greasing of the sheet bar is generally performed by roll application today. In particularly critical cases it may be necessary to use pressure lubrication in the tool itself.

9.6   SEAMS

Seaming of body sheet is a frequent forming operation in vehicle construction, and one which is particularly critical for some aluminium body sheet alloys. A glance at the forming limit diagram, Fig. 9.1, reveals that the deformation stress in seaming runs along the principal deformation axis ϕ_1 with $\phi_2 = 0$, where the forming limit curve is at a minimum. In addition, in the seam area, the drawn part has usually experienced the greatest initial deformation. The crystallographic characteristics of aluminium lead, precisely in conditions of high bending stress, to the formation of local shear zones or shear bands, in which the additional deformation concentrates (**16**)–(**19**) and which can lead to cleavage fracture. This is dependent on the type of alloy and the degree of initial deformation.

As a consequence of this behaviour, it is advisable, contrary to the practice of seaming steel body sheet using tools, to distribute the deformation as uniformly as possible in the bending zone, so that the bending radius is maintained.

Such a measure was proposed by Wolff (**14**) and leads to the formation of a so-called 'rope hem', which is illustrated in Fig. 9.11. Table 1.6 in Chapter 1 indicates which type of seam is recommended for various aluminium materials.

9.7   RIBS AND BEADS

Stretch forming is the predominant forming process for producing ribs and beads. Deep drawing, i.e. post-flowing of the material, facilitates the manufacture of larger ribs, but this generally requires relief blanks, in the

Bend/fold

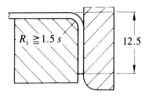

$R_i \geqq 1.5\,s$ 12.5

Tilt

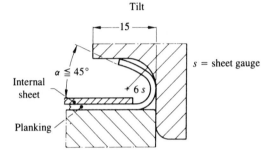

15

$\alpha \leqq 45°$

Internal
sheet

Planking

6 s

s = sheet gauge

Press

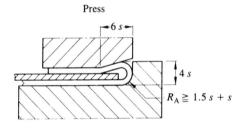

6 s

4 s

$R_A \geqq 1.5\,s + s$

**Fig. 9.11   Tool design for manufacturing a so-called 'rope hem', by N.P. Wolff
and Mercedes-Benz AG**

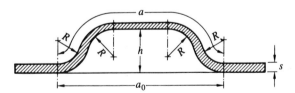

a

R R h R R

s

a_0

**Fig. 9.12   Minimum recommended radii for the manufacture of stretch-formed
ribs (14). Sheet thickness s, rib height h, radius *R*, and rib width, a_0.
Maximum rib height h_{max} is obtained from the stretched layout $s \leq$
1.25 · a_0. Recommended minimum radii: AlMg5Mnw–R = 10s; 6009–
T4–R = 12 s; 6010–T4–R = 15 s**

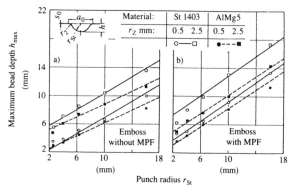

Fig. 9.13 **Maximum bead depth h_{max} as a function of punch radius r_{st} for closed semicircular beads without (*left*) and with material post-flow (*right*). Materials: St1403 and AlMg5(Mn), drawing edges $r_z = 0.5$ or 2.5 mm (20).**

corner areas of which the risk of crack formation must be taken into account. Figure 9.12 gives minimum recommended radii for ribs produced exclusively by stretch forming (**14**).

The embossing of closed semicircular beads has been extensively studied by Widmann (**20**) for St1403 and AlMg5Mn. Figure 9.13 illustrates some results of this study. An approximation formula for calculating the maximum bead depth of closed and open semicircular beads is as follows:

$$h_{max} = C \cdot n \cdot a_o \tag{11}$$

where

h_{max} = maximum bead depth

n = hardening exponent

a_o = die width, see Fig. 9.13

C = material constant

For the material AlMg5Mn, which is primarily used for ribbed interior parts of bodies, the material constant C is indicated as 0.8 for closed and 1.1 for open beads (**20**). At the end of the bead, a trapezoidal finish is preferrable to a radius-shaped one (**20**).

References

Chapter 1

(1) **Zengen, K. H. v.**, Aluminium in vehicle construction–the light alternative (in German), 'Praxis-Forum' series, *Fachbroschre Oberflächentechnik*, 1988, 12.

(2) **Baron, J. J.**, The aluminium body of the new Dyna (in German). *Aluminium*, 1954, **30**, (5), 183–194.

(3) **Takasuga, S.**, The transition to aluminium in the Japanese automobile industry and material performance requirements: focusing on wrought aluminium alloy, *International Symposium on Aluminium in Transportation: prospects for the Future*, 1989, Tokyo.

(4) **Tschopp, T.**, Ecological arguments for the use of aluminium in vehicles (in German), *Ingenieur-Werkstoffe*, 1991, **3**, (6), 15–17.

(5) **Glaser, K. F., and Johnson, G. E.**, *Construction experiience on aluminium experimental body*, 1974, SAE Paper No. 740075.

(6) **Burst, H. E., Buerle, H.-P., and Thull, W. F.**, *The all-aluminium auto body–a study based on the Porsche 929*, 1983, SAE Paper No. 830094.

(7) **Burst, H. E., and Thull, W. F.**, Concept and characteristics of an aluminium body for passenger cars (in German), *Aluminium*, 1986, **62**, (12), 915–923.

(8) **Hasler, F.**, The aluminium-intensive body (in German), Documentation volume, *VDI conference on Trends in modern body construction*, 1987, Wolfsburg.

(9) **Kewley, D.**, *The BL Technology ECV 2 energy conservation vehicle*, 1985, SAE Paper No. 850103.

(10) **Sheasby, P. G., and Wheeler, M. J.**, Aluminium structured vehicle technology: a review, *3rd International Symposium Aluminium and Automobil*, 1988, Documentation volume, Aluminium-Verlag, Düsseldorf.

(11) **Hoehl, K.**, The application of cast aluminium in the vehicle structure, based on the example of an urban car study (in German), *3rd International Symposium Aluminium and Automobil*, 1988, Documentation volume, Aluminium-Verlag, Düsseldorf.

(12) **Gausmann, W.**, Design and implementation of an aluminium composite structure for a sports car (in German), *3rd International Symposium, Aluminium and Automobil*, 1988, Documentation volume, Aluminium-Verlag, Düsseldorf.

(13) SAE Papers 870147 and 870148, Detroit, 1987.

(14) **Wheeler, M. J., Sheasby, P. G., and Kewley, D.**, *Aluminium structured vehicle technology–a comprehensive approach to vehicle design and manufacturing in aluminium*, 1987, SAE Technical Paper No. 870146.

(15) **Kewley, D., Campbell, I. G., and Wheatley, J. E.**, *Manufacturing feasibility of adhesively bonded aluminium for volume car production*, 1987, SAE Technical Paper No. 870150.

(16) **Nardini, I., and Seeds, S.**, *Structural design considerations for bonded aluminium structured vehicles*, 1989, SAE Technical Paper No. 890716.

(17) **Nardini, D., McGregor, I. J., and Seeds, S.**, *Analysis and testing of adhesively bonded aluminium structural components*, 1990, SAE Technical Paper No. 900795.

(18) **Warren, A.S., Wheatley, J.E., Marwick, W.F., and Meadows, D.J.**, *The building and test-track evaluation of an aluminium structured Bertone X1/9 replica vehicle*, 1989, SAE Technical Paper No. 890718.

(19) **Hasler, F., Munch, E., Reiter, K., and Timm, H.**, Automotive chassis, 1986, US Patent No. 4,618,163.

173

(20) **Evers, Th.,** Aluminium gaat steeds grotere rol spelen in autoindustrie, *Alurama*, 1988, **1**, 10.

(21) **McClure, R. H. G.,** *The aluminium intensive vehicle: a cost-effective lightweight structure for the automotive industry,* Autotech '89 (IMechE seminar paper).

(22) **Wehner, F., Rozczyn, H.-G., Gold, E. and Maier, J.,** Aluminium sheet and sections in the production car (in German), Ingenieur-Werkstoffe, 1990, **2**, (3), 37–40.

(23) **Uno, T., and Baba, T.,** Development of new aluminium alloy for auto body sheet, *Japan J. Light Metals*, 1978, **28** (4), 37–40.

(24) **Komatsubara, T., and Matsuo, M.,** *New Al–Mg–Cu alloys for autobody sheet applications,* 1989, SAE Technical paper No. 890712.

(25) **Ostermann, F.,** *Developments in the forming and working of sheet parts made of aluminium (in German),* Werkstatt und Betrier, 1988, (7/8), special publication by Aluminium-Zentrale e.V., Düsseldorf. The paper is a comprehensive summary of past developments.

Chapter 2

(1) **DIN 50919,** *Corrosion studies on contact corrosion in electrolyte solutions,* 1984 Edition (in German).

(2) **Strobl, C.,** Theory of corrosion, in *Korrosionsschutz bei Kraftfahrzeugen,* 1989, Haus der Technik, Vulkan-Verlag, Essen.

(3) **Godard, H. P.,** Aluminium, in *The corrosion of light metals,* 1967, John Wiley, New York.

(4) **Huppatz, W., and Wieser, D.,** The electrochemical behaviour of aluminium and opportunities for corrosion protection in practice (in German), *Werkstoffe und Korrosion,* 1989, **40**, 57–62.

(5) **Hatch, J. E.,** *Aluminium–properties and physical metallurgy,* 1984, ASM, p. 242.

(6) **Reiter, M.,** *Comparative corrosion tests on steel and aluminium materials* (in German), Diploma thesis, FH Berlin, March 1991.

(7) **Elze, J.,** Electrochemical series of metals ina practical corrosion medium (in German), *Werkstoffe und Korrosion,* 1959, **10**, 737–738.

(8) **Zeiger, H.,** Corrosion resistance of aluminium in contact with other metals (in German), *Aluminium,* 1961, **37**, 284–228.

(9) **Vosskhler, H., and Zeigler, H.,** The effect of alloy elements and other constituents on the corrosion performance of wrought aluminium materials (in German), *Aluminium,* 1961, **37**, 424–429.

(10) VG 81 249, Part 2, Corrosion of metals in sea water and a marine atmosphere, free corrosion in sea water (in German), Draft, May 1991.

(11) *Aluminium-Taschenbuch,* 1984, 14th Edition. Chapter 3, Cheical behaviour of aluminium (in German).

Chapter 3

(1) **Arit, K., and Brücken, T.,** Papers: *Developments in the surface treatment of Al* (in German). 1988, pp. 175–204.

(2) **Gold, E., Horn, W., and Maier, J.,** 1988, **42**, 248–253.

(3) **Roland, A. W.,** *Industrie Anzeiger,* 1988, **32**, 28–32.

(4) **Wittel, K.,** *I-Lack,* 1989, **57**, 97–102.

(5) **Treverton, J. A., and Davies, N. C.,** *Metals Technology,* 1977, 480–489.

Chapter 4

(1) **Burst, E., Buerle, H.-P., and Thull, W.,** Study of a monocoque all-aluminium body based on the Porsche 928 S (in German). *ATZ Automobiltechnische Zeitschrift,* 1984, **86**, No. 5.

(2) The use of high-strength steels in passenger cars (in German).
Part 1, **Drewes, E. J., Kraus, H., and Müschenborn, W.,** Materials and their characteristics, forming and joining properties.
Part 2, **Anselm., D., Henke, L., and Rauser, M.,** Strength, energy absorption and repair methods.
Part 3, **Krauss, H., and Roesens, D.,** Stiffness, vibration and acoustics. Special report in *ATZ Automobiltechnische Zeitschrift*, 1985, Nos 10, 11, and 12.

Chapter 5

(1) **Siegert, K.,** Examples of the use of aluminium in body construction (in German), *Symposium Aluminium and Automobil,* pp. 9/1–9/8, Documentation volume, 1981, Aluminium-Verlag, Düsseldorf.

(2) **Eichhorn, F., and Singh, S.,** *Surface treatment of aluminium parts as a preparation for resistance spot welding* (in German), Bänder-Bleche-Rohre, 1977, **11**, 499–505.

(3) *Resistance welding: measuring the transition resistance on aluminium materials* (in German), 1985, DVS-Merkblatt 2929.

(4) *Resistance spot and roller seam welding of aluminium and aluminium alloys of 0.35 to 3.5 mm individual gauge- preparation for and performance of welding process* (in German), 1986, DVS-Merkblatt 2932, T3.

(5) DIN 50124, *Shear tension test on resistance spot, resistance projection and fusion point welds* (in German), 1977.

(6) **Leuschen, B.,** The load-bearing behaviour of aluminium and aluminium/steel resistance spot welds under various loads (in German), 1984, Dissertation TH Aachen.

(7) **Eichhorn, F., Emonts, M., and Leuschen, B.,** Projection welding of deep drawing aluminium materials (in German), *Aluminium*, 1981, **57**, 9, 607–611.

(8) **Eichhorn, F., Emonts, M., and Leuschen, B.,** Projection welding of aluminium materials with different types of projections (in German), *Aluminium*, 1982, **58**, 8, 451–457.

(9) Resistance projection welding of aluminium materials of 0.35 to 3.5 mm individual gauge (in German), 1988, DVS-Merkblatt 2936.

(10) **Klock, H., and Schoer, H.,** Welding and brazing of aluminium materials (in German), *Fachbuchreihe Schweisstechnik*, **70**, 1977, Deutscher Verlag für Schweisstechnik (DVS) GmbH, Düsseldorf.

(11) **Leuschen, B.,** TIG d.c. and a.c. welding, with 'increaed electrical risk' (in German), *Trennen und Fügen* 1986, **17**, 13–16.

(12) **Behler, K., Beyer, E., and Schfer, R.,** Laser beam welding of aluminium (in German), *Aluminium*, 1989, **65**, (2), 169–174.

(13) **Beyer, E., Behler, K., Hoffman, K., and Berkmanns, J.,** Aluminium thin-sheet welding with the CO_2 high-output laser (in German), *VDI-Z*, 1991, **133**, (1), 82–86.

(14) **Hahn, O., Budde, L., and Boldt, M.,** *Joining by forming with and without an auxiliary joining part, compared with resistance spot welding* (in German), DVS Report 124, 1989, DVS-Verlag GmbH, Düsseldorf.

(15) **Hennings, J., and Maier, J.,** Characteristics of clinched joints in aluminium sheets compared with resistance spot welds (in German), *Aluminium*, 1983, **59**, (5), 358–363.

(16) **Maier, J., and Schröder, D.,** Strength characteristics of penetration joining bonds in aluminium sheet (in German), *DBF Colloquium Mechanical sheet joining techniques today*, 1991.

(17) **Steimmel, F.,** The effect of surface state and coatings on the strength properties of penetration-joined aluminium sheet (in German), *DFB Colloquium Mechanical sheet joining techniques today*, 1991.

(18) **Schmid, G., and Singh, S.,** Quality assurance of penetration joints in volume production (in German), *DFB Colloquium Mechanical sheet joining techniques today*, 1991.

Chapter 6

Books

(1) **Altenpohl, D.,** *Aluminium seen from inside* (in German), Aluminium-Verlag, Düsseldorf.
(2) Aluminium handbook (in German). Aluminium-Verlag Düsseldorf.
(3) **Laue, K., and Stenger, H.,** *Extrusion, processes–machines–tools* (in German), Aluminium-Verlag, Düsseldorf.
(4) *Aluminium key* (in German), Aluminium-Verlag, Düsseldorf.
(5) DIN Handbook 27, *Non-ferrous metals 2* (in German), Beuth-Verlag, Berlin/Cologne.
(6) Extrusion (in German), reports from Symposia of the Deutsche Gesellschaft für Metallkunde Oberursel.
(7) **Koewius, A., Gross, G., and Angehrn, G.,** *Aluminium design in commercial vehicle construction* (in German), Aluminium-Verlag, Düsseldorf.
(8) **Koser, J.,** *Designing with aluminium* (in German), Aluminium-Verlag, Düsseldorf.
(9) Text book *Aluminium materials technology for vehicle construction* (in German), Aluminium-Verlag, Düsseldorf (in preparation).

Journals

(10) *Aluminium,* Aluminium-Verlag, Düsseldorf.
(11) *Aluminium-Kurier,* Aluminium-Verlag, Düsseldorf.

Technical papers

(12) **Gitter, R.,** Designing with extruded aluminium sections (in German). Tagungsband DVM-Tag 1990, Berlin.
(13) **Maier, J., and Schrder, D.,** Penetration joints in Al: strength properties (in German). *Bänder–Bleche–Rohre,* 1991, **5**, 83.
(14) Reports of the *Symposium Aluminium and Automobil,* Aluminium-Verlag, Düsseldorf, 1980, 1988.
(15) *Aluminium und Automobil,* Aluminium-Verlag, Düsseldorf, 1989; 1991.

Standards

DIN 1712/3, *Aluminium semi-finished products, composition* (in German).
DIN 1725/1, *Aluminium wrought alloys, composition* (in German).
DIN 1732, *Welding fillers for aluminium* (in German).
DIN 1746, *Tubes, strength values, technical specification,* (in German).
DIN 1747, *Rods, strength values, technical specification,* (in German).
DIN 1748, *Sections, strength values, tech. specification, design, dimensional tolerances,* (in German).

Chapter 7

(1) **Sonsino, C. M.** *et al.,* Report of the Fraunhofer-Institut für Beitriebsfestigkeit, Darmstadt, 1988.
(2) **Shroyer, H. F.,** US Patent 2830 343 of 15.4.1958.
(3) **Wenk, L.,** *Giesserei,* 1989, **76**, (9), 655–672.
(4) **Medana, R.,** Current status and technology of full-mould casting of production parts (in German), 1988, VDG Seminar, Hilden.
(5) **Smetan, H.,** New process technologies for casting (in German), 18th Continuing Education Seminar for Die-casters, 1987, Düsseldorf.
(6) **Lismont, H.** *et al.,* PreCoCast and Cobapress (in German), Seminar *Aluminium mould castings using innovative technology,* 1990, Meschede.
(7) **Chadwick, G. A.** *et al.,* *Met. Mater.,* 1983, **3**, 38–42.
(8) **Barlow, J.,** IAVD Congress on Vehicle Design and Components, 1984.

(9) **Ziwi, Y.** *et al., Giesserei-Praxis*, 1983, **3**, 38–42.
(10) **Flemings, M. C.** *et al., AFS-Trans.*, 1973, 81–88.
(11) **Erz, H.-P.,** Paper presented at German Foundry conference, 1990, Bad Homburg.
(12) **Lange, B.,** Quenched and tempered aluminium die castings for safety parts (in German), Seminar *Aluminium mould castings using innovative technology*, 1990, Meschede.
(13) **Schneider, W.,** New materials and processes for innovative aluminium castings (in German), 1990, Meschede.
(14) Honda Deutschland GmbH press information, 1990 (in German).
(15) Company announcement Homet Corporation, in *Precis. Met.*, 1987, **45**, (12), 20.
(16) **Clopham, L., and Smith, R.,** *Cast Metals*, 1989, **2**, 11–15.
(17) **Warode, A.,** Diploma thesis, 1990, FH Dortmund.
(18) **Fuchs, H. A.** *et al.,* Composite materials with ceramic fibres in a light alloy matrix (in German), *Werkstoff und Innovation*, 1988, **2**, 42–47.
(19) **Rudy Hoffman, J.,** Integrated processing of aluminium casting incorporating continuous aluminium fibres, ASM Conference, 1990, Amsterdam.
(20) **Dinwoodie, J.** *et al., Proceedings of the 5th International Conference on Composite Materials*, 1985, San Diego, CA.
(21) **Folgar, E.** *et al., Proceedings of the 8th Annual Conference on Compositers and Advanced Materials*, 1984, American Ceramic Society, Cocoa Beach, FL.

Chapter 8

(1) Designing with light alloy forgings (in German), *Werkstatt und Betrieb*, 1979, **112**, (4).
(2) Design guidelines for drop forged parts made of aluminium alloys, *Aluminium*, 1940, 198–204.
(3) **Ostermann, F., and Söllner, G.,** Solid formed parts of aluminium in vehicle construction (in German), *7th International light alloy meeting*, 1981, Leoben/Vienna, pp. 362–364.
(4) *Aluminium forgings* (in German), Report 32 of the Aluminium-Zentrale.
(5) **DIN 1749,** Parts 1–4.
(6) Vehicle bodies and chassis, lightweight design and engines and power transmission (in German), *VDI-Jahrestagung*, 1978.
(7) *Aluminium handbook* (in German), 14th ed., 1988.
(8) **Lowak, H., Grubisic, V.,** Methods of forecasting service life for components made of aluminium alloys (in German), *Materialprüfung*, 1985, **27**, (11), 337–343.

Chapter 9

(1) **Siegert, K.,** Comparison of body sheet made of aluminium and steel (in German), *Aluminium* 1983, **59**, (5/6), 363–366 and 438–442.
(2) **Hasek, V., and Werle, T.,** *Deformation analysis by means of the line network process* (in German), Umformtechnik, publisher K. Lange, Vol. 3, Sheet working, Ch. 3, 1990, 2nd ed. Springer-Verlag.
(3) **Schmoeckel, D., and Heller, C.,** *Forming aluminium sheet at high temperatures* (in German), Research report no. 34 (1988), Deutsche Forschungsgesellschaft für Blechverarbeitung.
(4) **Blaich, M.,** *The drawing of sheet parts made of aluminium alloys* (in German), Reports from the Institut für Umformtechnik, University of Stuttgart, No. 61, 1981, Springer-Verlag.
(5) **Kostron, H.,** The mathematics of the tensile test (in German), Archiv Eisenhttenwesen, **22**, 1951, 9/10, 317–325.
(6) **Reissner, J., Schmid, W., and Meier, M.** Deep drawing (in German), Umformtechnik, Publisher K. Lange, Vol. 3, Sheet working, Ch. 7, 1990, 2nd ed, Springer-Verlag.

(7)   *Aluminium handbook* (in German), 14th ed., 1984, Aluminium-Zentrale e.V., Düsseldorf, p. 93.

(8)   **Breidohr, B.,** *Studies on the deep drawing of parts with uneven base shapes* (in German), DFB Research report, 1988, Deutsche Forschungsgesellschaft für Blechbearbeitung e.V.

(9)   **Ostermann, F.,** Developments in the forming and working of sheet parts made of aluminium (in German), Werkstatt und Betrieb, 1988, **121**, (7/8), 545–550 and 623–625.

(10)   **Witthser, K.-P.,** Examination of test methods for assessing friction conditions in deep drawing (in German), Dissertation, TU Hanover, 1980.

(11)   **Balbach, R.,** *Optimisation of surface microgeometry of aluminium sheet for body drawing* (in German), Report no. 98, Institut für Umformtechnik, Universität Stuttgart, 1988, Springer-Verlag.

(12)   **Mössle, E.,** *The effect of the sheet surface in drawing of sheet parts made of aluminium alloys* (in German), Report no. 72, Institut fur Umformtechnik, Universität Stuttgart, 1983, Springer-Verlag.

(13)   **Woska, R.,** The effect of selected surface coatings on friction and wear in deep drawing (in German), Dissertation Institut fur Umformtechnik, T.H. Darmstadt, 1982.

(14)   **Wolff, N. P.,** *Interrelation between part and die design for aluminium auto body panels,* 1978, SAE Technical Paper, No. 780392.

(15)   **Siegert, K., and Thoms, V.,** The use of lubricants to influence friction during forming of body sheet (in German), *Aluminium,* 1987, **63**, (4), 401/409.

(16)   **Akeret, R.,** Failure mechanisms in the bending of aluminium sheet and the limits of bending (in German), *Aluminium,* 1978, **54**, 117–123.

(17)   **Akeret, R.,** The failure of aluminium materials in forming due to localised shear zones (in German), *Aluminium,* 1978, **54**, 193–198.

(18)   **Akeret, R.,** Observations of the localisation of deformation in aluminium materials (in German), *Aluminium,* 1978, **54**, 385–391.

(19)   **Akeret, R.,** The forming properties and structure of wrought aluminium materials (in German), *Aluminium,* 1990, **54**, 147–150 and 246–252.

(20)   **Widmann, M.,** Beading during drawing of large sheet parts (in German), DFB Research report no. 21, *Stahl und Eisen,* 1984, **104**, (4), 187–192.

INDEX

182   *Index*